NOT IN A THOUSAND YEARS

The Uniqueness of the 20th Century

NOT IN A THOUSAND YEARS
The Uniqueness of the 20th Century

Events | Inventions | Attainments
Revelations | Phenomena | Chain Reactions

Geoff Fernald

Fernald, Geoff, *Not In A Thousand Years*

13 Digit ISBN   978-0-9789354-0-5
10 Digit Traditional ISBN   0-9789354-0-3
Keywords: Cultural history, twentieth century, turning points, social change, civilization, inventions, mobility, communication, Earth, Universe, health, human biology,
the arts, electric power, energy, computer aided design, cognitive sciences, and twenty-
first century.

To order (email preferred):
books@fernald.org

Geoff Fernald
PO Box 1052
San Carlos, California 94070-1052
Telephone 603-569-1538

To
Kenneth Clark for his presentation
on public television of *Civilization*
and
Castle Fernald for his work
on the first jet engine in America

"Ozymandias"

I met a traveler from an antique land
Who said: — Two vast and trunkless legs of stone
Stand in the desert... near them, on the sand,
Half sunk, a shatter'd visage lies, whose frown,
And wrinkled lip, and sneer of cold command,
Tell that its sculptor well those passions read
Which yet survive, stamp'd on these lifeless things,
The hand that mock'd them and the heart that fed;
And on the pedestal these words appear —
"My name is Ozymandias, king of kings:
Look on my works, ye mighty, and despair!"
Nothing beside remains: round the decay
Of that colossal wreck, boundless and bare
The lone and level sands stretch far away.

Percy Bysshe Shelley 1818
(1792 — 1822)

Babel.
This then is the premise:
There are large forces at work
Underneath our daily lives
That begin to hint similarities in
Other centuries to pivotal times and events:
Alexandria's bibliotheca,
Socrates' gymnasium,
Buddha's middle way,
Christ's dying for all of us,
Mohammed's engagement of
Desert nomads,
The plague,
Renaissance art and Copernican science,
Mozart's and Bach's high classical,
Gutenberg's press,
Industry's machines,
This twentieth century of ours begins the undoing of Babel
A final emergence from miscommunication...
We now find ourselves
Leaping to our assignment from God
who commanded —
"Copy me."

Geoff Fernald 2003

ACKNOWLEDGEMENTS

For contributions to the list of subjects I thank Nelson Wright, Chris Cole, Hugh Larsen, Toby Fernald, Jeff Newell and the many others who made encouraging suggestions; and for most of the detail work on that list, Wendy Fong. For his many readings, suggestions, ideas, criticisms and rewrites, I thank Kevin Bergeron. For technical accuracy and an unflinching eye for quality John van Raalte. For extensive help with grammar and her general encouragement to continue I give my sincere appreciation to Dr. Gail Bickford of Freedom Press. For the Life Utilities chapter in all respects I thank Jim Ragano. For the Arts chapter I thank Toby Fernald and Charlie Kegel. For their readings and suggestions, the help of: Don McPherson, Toby and Beau Fernald, Bret Stephenson, Wendy Fong, Judy Crosley and Kay Baliones. For cover design Eli Fernald and Brad Marion. And to Jane my deepest appreciation for typesetting and layout design. All of the book's weaknesses are mine alone. The 20th Century deserves the credit for what we accomplished with its thousands of unheralded contributors.

CONTENTS

Introduction

Our search for meaning has been and continues to be a central focus of our hearts and lives. Is meaning to be found in family, in work and contribution, in study and learning or in faith in a higher power? Are our lives really as small as they feel, as spoken by Shelley's poem? Even for the king of kings, is there eventually no memory or substance? These few pages and documenting tables reveal that we, participants of the 20th Century, more than ever before are able to see ourselves in our role in the creation of social change with perspective. That perspective adds meaning to our lives. We become part of the riverbed through which these changes flow as we form new paths in life on this industrialized earth. Each of us becomes part of its inexorability — from the guy pumping gas at the Shell gas station in the '50s to the female Navy lieutenant who invented software, on up through the greatness of Albert Einstein and Dr. Fleming — each of us is included.

In late 1998 I began to collect lists of the major impact accomplishments from the 20th Century. I did this in response to a TV show, which ostensibly did the same. During that show's listing I became so angered by the media-centric shallowness of their list that I embarked on a journey to create my own summary, hoping for depth and durability. As I progressed, I inveigled time with anyone who would talk to me — over lunch, dinner, in the hallway — to help me create a list of contributions that had significant and universal impact on multiple millions of lives.

So began my personal search for relevance in my life's century. Together you and I will walk through the lists by category, starting with the largest and moving to the smallest grouping of major contributions, examining them for perspective and extent. You will be astounded by the names and achievements taken in total, and by their essential role in creating only the tip of the iceberg of contribution in our century. Thereafter you are on your own to read the lists, add to them, and make your own interpretation; delete those you feel are over-stated, continuing to assess what really happened in the 20th Century. At the end we will speculate on what is ahead in the biology century we have now begun. The 20th Century was definitely the electronic century as the 15th was the fine art century and the 18th the classical music century.

This story is intended to be worldwide in scope, but will by its nature, being written in the United States, be narrow, awaiting addition and reorganization by other major social masses.

The following chapters walk you through this overall list of the more than 350 discoveries, inventions and events of greatest impact during the 20th Century. They are combined and listed at the end of the book for easy reference. In some cases the inventor is named, in others the major implementer, depend-ing on the differences in their timing and contribution. The list will contain some things that are astounding and others that are obscure. Each item has been placed in a category that makes assessing its overall impact easier. The categories were chosen from several classical historical studies, later referenced, and partly by natural groupings driven by this century's life itself.[1] These categories were then organized by type and frequency-of-list-occurrence, and are shown in or-der from largest influence to smallest on the next page. The category summary is followed by an abbreviated Glimpse List of the more recognizable 182 of the full century list. The book is organized into chapters whose order proceeds from largest category to smallest. This structure does not translate directly into most important to least. I hope reading through these chapters leads you, too, on a journey of personal discovery.

Geoff Fernald

Reference Note
1 See the bibliography at the end of the book.

QUANTIFYING CHANGE EVENT FREQUENCY BY CATEGORY

Chapter	Category	Count
1	Social Change	57
2	Mobility and Communication	57
3	Physical Knowledge of our Universe	30
	Earth History	15
4	Life Utilities	31
5	Health	29
	Understanding our Body	15
6	Fine Art, Music, Dance	26
	Poetry, Literature	12
	Recording of Life Past, Present and Future	13
	Humor	4
	Fantasy	1
7	Human Psychology	12
	Philosophy	6
	Technical Uncertainty	6
	Learning, Teaching and New Tools	11
8	Dwellings and Structures	5
	Government	3
	War	17

A Glimpse of 20th Century Accomplishment

CHANGE EVENTS	YEAR	CHANGE AGENT	IMPACT CATEGORY
Interpretation of Dreams	1900	Freud	Psychology
Quantum Theory	1900	Planck	Knowledge of Universe
23 Open Math Questions	1900	David Hilbert	Technical Uncertainty
Gyroscope (MIT)	1901	Draper	Mobility/Communication
X-rays	1901	Roentgen	Understanding the Body
Air Conditioning	1902	Carrier	Life Utilities
B&W Photography as Art	1902	Stieglitz	Recording Life
First Electric Utility	1903	Edison	Life Utilities
Powered Flight	1903	Wright	Mobility/Communication
Vacuum Tube for Radio	1904	Fleming	Mobility/Communication
Special Relativity E=mC^2	1905	Einstein	Knowledge of Universe
IQ test	1905	Binet/Simon	Psychology
First Women's Vote	1906	Finland	Social Change
Radioactive Dating	1907	Boltwood	Earth History
RF Amplifier	1908	de Forest	Mobility/Communication
Automobile Assembly Line	1908	Ford	Mobility/Communication
Bakelite (first plastic)	1910	Baekeland	Life Utilities
China Abolishes Slavery	1910	China	Social Change
Inside the Mind	1911	Adler	Psychology
Superconductivity	1911	Onnes	Knowledge of Universe
Atomic Nucleus	1911	Rutherford	Knowledge of Universe
Oil Cracking for Gasoline	1912	Burton	Mobility/Communication
Sinking of Titanic	1912	SS Titanic	Technical Uncertainty
Subatomic Behavior	1913	Bohr	Knowledge of Universe

CHANGE EVENTS	YEAR	CHANGE AGENT	IMPACT CATEGORY
Home Refrigerator	1913	Wolf/Mellowes	Life Utilities
NY Show — modern art	1913	Kuhn	Fine Art, Music & Dance
Income Tax	1913	Congress	Government
Blood Bank	1914	Hustin	Health
Proton	1914	Rutherford	Knowledge of Universe
Behavioral Psychology	1914	Watson	Psychology
Panama Canal	1914	USA	Dwellings and Structures
WWI End of Colonialism	1914	Germany	War
Movie Comedy	1915	Chaplin	Recording Life
The Trial	1915	Kafka	Government
First Birth Control Clinic	1915	Sanger	Social Change
Continental Drift	1915	Wegener	Earth History
Bolshevik Revolt	1917	Lenin	Social Change
Movies: MGM	1917	Meyer	Recording Life
First Congresswoman	1917	Rankin	Social Change
Millions of influenza deaths	1918	Spanish Flu	Health
Women's Vote	1920	Anthony	Social Change
Prohibition	1920	Amendment	Social Change
Civil Disobedience	1920	Gandhi	Social Change
Insulin for diabetics	1922	Banting	Health
The Skyscrapers	1923	Corbusier	Dwellings and Structures
Frozen Fish Process	1923	Birdseye	Life Utilities
Modernized Turkey	1923	Ataturk	Social Change
Wave/Particle Duality	1923	de Broglie	Knowledge of Universe
TV camera tube	1923	Zworykin	Mobility/Communication
Universe>Milky Way	1924	Hubbel	Knowledge of Universe
The Loud Speaker	1924	Kellog	Mobility/Communication
Rocket	1926	Goddard	Mobility/Communication
Xray CAT Scans	1926	Ledley	Understanding the Body
Real Sound Movies	1927	Warner	Mobility/Communication
Uncertainty Physics	1927	Born	Knowledge of Universe
Father of the Blues	1927	House	Fine Art, Music & Dance
PVC Modern Plastic	1927	Klatte	Life Utilities
Quartz Clock	1927	Marrison	Mobility/Communication
Sports Hero Model	1927	Ruth	Social Change
Color Film	1928	Eastman	Recording Life

CHANGE EVENTS	YEAR	CHANGE AGENT	IMPACT CATEGORY
Penicillin discovered	1928	Fleming	Health
Musical Drama	1928	Gershwin	Fine Art, Music & Dance
Magnetic wire recording	1928	Begun,Pfleumer	Mobility/Communication
Scotch Tape	1929	Drew	Life Utilities
Revolutionize auto transport	1929	Autobahn	Mobility/Communication
Child Development Phases	1929	Piaget	Psychology
Stock Market Crash	1929	USA	Social Change
First Electrical Computer	1930	Bush	Mobility/Communication
CFC Invented	1930	Midgley	Life Utilities
Ideal Family Paintings	1930	Rockwell	Recording Life
Math is Uncertain — Proof	1931	Goedel	Technical Uncertainty
Brave New World	1932	Huxley	Recording Life
Polaroid film	1932	Land	Recording Life
Welfare, The New Deal	1933	Roosevelt	Social Change
Prohibition Repealed	1933	USA	Social Change
"Falsifiability" of theories	1934	Karl Popper	Technical Uncertainty
Sulfa Drugs	1935	Domagk	Health
Polling (statistical sampling)	1935	Gallup	Social Change
Airlines — Pan American	1935	Trip	Mobility/Communication
Alcoholics Anonymous	1935	Wilson	Social Change
Major builder of The Blues	1936	Johnson	Fine Art, Music & Dance
Deficit Economics	1936	Keynes	Social Change
Capitalism/Organized Labor	1936	Reuther	Social Change
Nylon-synthetic fabrics	1937	Carothers	Life Utilities
American Folk Music	1937	Guthrie	Fine Art, Music & Dance
American Realism	1937	Wyeth	Fine Art, Music & Dance
Ball Point Pen	1938	Biro	Life Utilities
Jazz vocal recordings	1938	Fitzgerald	Fine Art, Music & Dance
Skinner Box	1938	Skinner	Psychology
Nuclear Electric Power	1939	Fermi	Life Utilities
WWII triggered	1939	Hitler	War
DDT	1939	Muller	Life Utilities
Television	1939	Sarnoff	Recording Life
First Jet Engine Aircraft	1941	Whittle	Mobility/Communication
L' Etranger (The Stranger)	1942	Camus	Philosophy
Duck(duct) Tape	1942	Johnson	Life Utilities

CHANGE EVENTS	YEAR	CHANGE AGENT	IMPACT CATEGORY
Spread Spectrum idea	1942	Lamarr, Antheil	Mobility/Communication
Existentialism	1943	Camus	Philosophy
Pap Smear Test	1943	Papanicolaou	Health
DNA is genetic material	1944	Avery	Understanding the Body
Mathematical Encryption	1945	Bletchley	Mobility/Communication
Penicillin produced	1945	Florey	Health
Software	1945	Hopper	Learn and Teach Tools
Marshall Plan	1945	Marshall	War
Atomic Bomb — Hiroshima	1945	Oppenheimer	War
United Nations	1945	Wilson	War
ENIAC Computer	1946	Presper Jr.	Learn and Teach Tools
MRI	1946	Purcell	Understanding the Body
Baby and Child Care	1946	Spock	Social Change
Auto Smog — Los Angeles	1946	Tucker	Earth History
African American Sports	1947	Robinson	Social Change
The LP vinyl record	1948	Goldmark	Fine Art, Music & Dance
The Transistor	1948	Shockley	Mobility/Communication
1984	1949	Orwell	Learn and Teach Tools
Credit Card	1950	Diners	Mobility/Communication
Salk Vaccine for polio	1952	Salk	Health
Sedatives	1952	Wilkins	Life Utilities
Color TV	1953	RCA	Mobility/Communication
DNA double helix structure	1953	Watson, Crick	Understanding the Body
Sub-Four Minute Mile	1954	Bannister	Understanding the Body
Kidney Transplant	1954	Murray	Health
Rock and Roll	1954	Presley	Fine Art, Music & Dance
Nuclear Submarine (first strike threat ended)	1954	Rickover	War
Tetracycline antibiotic	1955	Conover	Health
Disneyland	1955	Disney	Fantasy
Green Eggs & Ham	1955	Geisel	Learn and Teach Tools
McDonalds — Fast Food	1955	Kroc	Social Change
End to Racism	1955	Parks	Social Change
Composite Carbon-fiber	1955	Rolls	Life Utilities
Global Interconnection	1956	Airfreight	Mobility/Communication
Video Tape	1956	Dolby	Mobility/Communication

CHANGE EVENTS	YEAR	CHANGE AGENT	IMPACT CATEGORY
Container Ship — Ideal-X	1956	Mclean	Mobility/Communication
Radio Astronomy	1957	Lovell	Knowledge of Universe
Objectivism — Individual	1957	Rand	Philosophy
First Satellite — Sputnik	1957	USSR	Knowledge of Universe
Xerox Dry Copying	1958	Carlson	Mobility/Communication
Laser/Maser	1958	Gould	Mobility/Communication
Implantable Pacemaker	1958	Greatbatch	Health
Marxism — Chinese Order	1958	Zedong	Social Change
Integrated Circuit	1959	Noyce/Kilby	Mobility/Communication
Plate Tectonics	1960	Ewing	Earth History
OPEC Oil Price Control	1960	OPEC	Social Change
Oral Contraceptive	1960	Pincus	Social Change
First Manned Satellite	1961	Gagarin	Mobility/Communication
Peace Corps of America	1961	Kennedy	Social Change
Homo Habilus	1961	Leakey	Earth History
DNA sequence Decoded: DNA=>RNA=Protein	1961	Nirenberg	Understanding the Body
Moon Rocket	1961	USSR	Knowledge of Universe
Silent Spring	1962	Carson	Earth History
Light Emitting Diode(LED)	1962	Holonyak	Mobility/Communication
First Communications Satellite	1962	NASA	Mobility/Communication
Rock and Roll world wide	1963	Beatles	Mobility/Communication
Folk Music Re-Invented	1963	Dylan	Social Change
Feminine Mystique	1963	Frieden	Social Change
"I Have a Dream " speech	1963	King Jr.	War
Medicare Act of 1964	1964	Johnson	Health
Vietnam War — Guerilla War	1964	Johnson	War
Black Power Awareness	1964	Malcolm X	Social Change
Social Sci-Fi Philosopher	1966	Roddenberry	Social Change
Heart Transplant	1967	Barnard	Health
Hippies Summer of Love	1967	Ashbury	Social Change
Acid Rain	1967	Coal	Earth History
Green Revolution — grain	1968	Borlaug	Earth History
2001: A Space Odyssey	1968	Clark	Poetry and Literature

CHANGE EVENTS	YEAR	CHANGE AGENT	IMPACT CATEGORY
Muppets — Educators	1969	Henson	Learn and Teach Tools
Powerful Woman Politician	1969	Meir	Social Change
First Person on the Moon	1969	USA	Knowledge of Universe
Fiber Optic telephony	1970	AT&T	Mobility/Communication
Public Keys, Encryption	1970	Ellis	Mobility/Communication
Cleaner Exhaust Cars	1970	EPA	Earth History
Pocket Calculator	1971	Texas Instruments	Life Utilities
First Mars Rocket Landing	1971	USSR	Knowledge of Universe
Roe v. Wade — Abortion	1973	US Courts	Social Change
Ozone layer and Freon	1974	Molina	Earth History
Genocide	1975	Pot	War
Cell Phone	1977	Cooper	Mobility/Communication
Apple II	1977	Wozniak, Jobs	Mobility/Communication
Future Revealing Movies	1977	Lucas	Social Change
CAT scanning	1979	Hounsfield	Understanding the Body
Inexpensive CD Music	1979	Sinjoou, Doi	Mobility/Communication
CAD (computer aided design)	1982	Multiple Contributors	Life Utilities
HIV and AIDS Epidemic	1983	Mutation	Social Change
MacIntosh Computer	1984	Jobs, Wozniak	Learn and Teach Tools
Fast Mail	1985	Fedex	Mobility/Communication
Nuclear Reactor disaster	1986	USSR	Life Utilities
Financial Failure of Marxism	1989	USSR	Social Change
World Wide Web (WWW)	1989	WWW	Mobility/Communication
Self Help Book Mania	1990	Boomers	Psychology
Human Genome Project	1991	Venter-Celera	Understanding the Body
Realistic TV drama	1992	Bochco	Recording Life
Integration begins in Africa	1994	Mandela	Social Change

CHAPTER 0

Winter on Fifth Avenue, New York City, Photo by Alfred Stieglitz, 1893
National Gallery of Art, Washington, D.C., Alfred Stieglitz Collection

1899 — The Beginning of the 20th Century

If you are under thirty you may not be familiar with the beginnings of the 20th Century. And that is as it should be. Renewal is one of the greatest advantages of the human cycle, shaking off the old and seeking the new. To understand the significance of *Not in a Thousand Years*, the events about to be described, we return briefly to what life was like in the Eastern United States at the turn of the century, which best represented a nation on the brink of the 20th Century. Europe was, at that time, also a center of great innovation and influence. London's coal haze in winter is often described in stories of that time as are the great philosophers in France and Germany. Innovation poured forth in these countries as many of the names testify; Diesel, Bessemer, Faraday and Maxwell. I beg your indulgence in my use of the United States as the example for the entry into the 20th Century. It provides the most easily described country for identifying a standard of the time, offering the reader a snapshot of life at the opening of the 20th Century.

Realizing that only thirty-three of the eventual fifty states were part of the Union of the United States in 1860, we see that much changed just prior to 1900. Most of the Southwest and Far West were added through war and purchase by century's end. The United States entered 1900 with its "Manifest Destiny" of coast to coast boundaries complete.

It is December of 1899, Lincoln's Emancipation Proclamation ending the Civil War had only decades earlier declared the Negro equal and free. And as with all social change, in spite of that declaration, significant time would pass (extending even to the end of the following century) before Lincoln's words were made true. Immigration in 1899 was uncontrolled, though first restrictions were beginning to emerge — Chinese immigrants were not allowed to become citizens following a declaration by Congress in 1882. A quota system followed in 1924 which set immigration limits based on both nationality and each nationality's existing proportion of population at that time in the United States. These immigration limits were typical of a nationalistic trend in Western behaviors during the closing decade of the 19th Century. The 19th Century's recent extension of the right to vote to all United States males, land owning or not, was becoming more and more the norm in other countries.

Late in the 19th century, East Coast Americans felt their easy access to places in their own country improving, with steam powered trains running on railroads that stretched all the way to California. A cross-country trip took less than four days coast to coast. Steamships regularly plied the Atlantic and Pacific oceans, providing access to Asia's and Europe's historic riches and ancient culture. Travel to America for Europeans was becoming more frequent. Commercial shipping flourished worldwide, though horse and buggy were still the local means of travel. The bicycle was available after 1880 and provided more accessible mobility.

Paper and package mail kept people in touch, though with a tendency for long delays from coast to coast, until the telegraph and railroad postal service began in 1869. Families tended to live out their lives in clustered adjacency. People in Los Angeles did not know that their state, California, had been admitted to the Union until six weeks after the fact. Mail often took months.

Only those lusting for adventure, new life and free land were willing to travel from coast to coast through unsettled territories. Their horse and oxen-drawn wagon trains rolled across the heaving land, westward.

Immigration into the United States was, as mentioned earlier, largely uncontrolled and light, until very late in the century. The Irish, driven by the potato famine, the Germans and Italians seeking less despotic regionalism, moved across the oceans and the face of America to find better lives. Sometimes they eked out a spare living in ports of entry and in the ghettos that rose around them where they were exploited in unrestricted labor markets. Classlessness was not yet abundant, in spite of the vaunted freedoms of America. There were still many levels; well-to-do educated whites, poor immigrants in their ethnic ghettos, women, and the Negro.

Daily life involved grocery shopping each and every day for perishables. Food preservation by refrigeration was non-existent. Ice coolers and cold weather acted only to slow food spoilage in winter. Basements were used to keep potatoes and various root vegetables cool. Vegetables from California or Florida were not available in the East. Small local farms provided the only fresh vegetables, milk and meat while "mom and pop" stores provided nearly everything that was for sale. No malls, no chain grocery stores. Chain department stores, such as F.W. Woolworth's Five & Dime, only emerged in large cities late in this pre-20th Century. Access to variety in products depended completely on one's geographical and urban location.

The first electric utility was built on Manhattan in 1882. Lighting was mostly by wax candle and oil lamp. Daylight was a precious commodity.

In Europe, science proceeded with a number of significant individual contributions such as Faraday's and Gauss's elaboration of electrical and magnetic fields, Maxwell's far reaching electromagnetic equations, Mendel's investigations of heredity with sweet pea plants and Darwin's publication of *The Origin of the Species*. Pasteur found a way to sterilize wine, vinegar, beer and eventually milk, allowing transport of these over longer time frames and distances without spoilage. And Bessemer invented a way to improve the production of high strength steel, one step toward the coming construction of tall buildings in the 20th Century.

In fact critical pieces of the groundwork were laid for the thousands of constructions, products and processes that would be invented and developed in the 20th Century. Nineteenth century inventions such as the lead-acid battery, dynamite, the Daguerreotype for photography, various motors — both electric and oil fuel based, were all invented and produced in small quantities late in the 19th Century. Textiles benefited greatly from the inventions of the cotton gin late in the 18th Century and the powered loom in the mid-nineteenth Century. The amount of wheat produced and the cost of grain improved greatly as a result of the invention and production of the horse drawn reaper (later called the combine) by McCormick.

Modern germ/virus/bacterial understanding of health and of our bodies began with a trickle of innovations in the 19th Century that would become a torrent in the 20th Century. John Snow empirically verified the idea that germs from one person spread disease to others during cholera epidemics in England and published his ideas in 1849. Other discoveries followed — pain reduction during surgery using ether as an anesthesia, and the reduction of infection during recovery from surgery by using carbolic acid as a sterilizing agent. Wilhelm Wundt and Franz Brentano introduced early ideas about human behavior, not yet called psychology, in the final forty years of the 19th century. Their use of empirical data to validate or invalidate such ideas was a scientific innovation. Statistical mathematics was developed to assist in the use of such data. The application of scientific method to fields previously thought more metaphysical began to anchor the health sciences in logic. Mendeleyev created the periodic table of the elements revealing the underlying order of all fundamental chemical components in the universe.

The arts flourished in major ways giving us Impressionism and its

The combine changed the rate of food harvesting. (Massey Fergeson 9790 above)

many painterly offshoots as well as the music of Beethoven and the fanta-
sy of Art Nouveau. English poetry and literature had one of their greatest
eras since Shakespeare with the introduction of the novel: Austen, Blake,
Dumas, Pushkin, Balzac, Dickens, Alcott, Tolstoy, Thoreau, Trollope,
Twain, Hardy, James, Poe, Kipling, Wilde, Wordsworth, the Brownings,
Wharton, Cather, Frost, Victor Hugo, Flaubert, Goethe, Schiller, etc...
Jules Verne's book *20,000 Leagues Under the Sea* ushered in the era of science fiction for the 20th Century. Walt Whitman and Emily Dickenson
changed the face of poetry.

Religion, which formed the great social fabric of people's lives in this
world before mass entertainment and automobiles, still dominated world
culture during the 19th Century. Sunday was a day of closed stores in
the United States in the late 1800s. It was God's day of rest for Christians. There was no radio or television available to intervene. Baseball
games and churches were well attended. Otherwise, strolls along park
lanes and visits with family or neighborhood gatherings were the fare of
the weekend. No U.S. president had been elected who did not profess
to be a Christian. Speaking further of religion, Joseph Smith invented
a new one, Mormonism, and American Christian evangelists joined in
the spread of colonial Europe and their promise to fill the world with the
"true faith"— Christianity; focusing on Africa and Asia. The developed
Western world was a world of Christianity in late 1899. Other faiths,
Mohammedism, Hinduism, and Buddhism remained strong on the continents of Africa and Asia and in the near-east.

For the United States of the early 1900s, the Christian religion dominated the population with 73 million adherents out of a population of 75
million people. People of that era would be surprised to learn that by the
year 2000 there would be a population of 278 million and an amazing 30
million non-believers! Only 44 percent of that coming population would
continue to attend church regularly. In that same future Muslim and Jew-
ish believers grew to 4 and 5.6 million, respectively, from almost nothing
in 1899. The Christian faith remained strong at 235 million. Of those
235 million, Evangelicals (non-traditional Christian sects) grew to an
unprecedented 138 million! Independent sect Christians went from 6
million in 1900 to 78 million by 2000. Catholics, who began the century
with only 1/3 the total Christian faithful (2/3 being protestant), would
swell to equal the Protestant population at 60 million by the year 2000.

Philosophy gained legendary contributors — Thomas Jefferson, Karl Marx, Schopenhauer, Nietzsche, Hegel, Kierkegaard, and John Stuart Mill. Lincoln and Napoleon led with a new vision of organized government.

The American college system flourished into a 2000 strong college and university system spanning the United States.[1]

And of course there was war; the world was in its final throes of colonial land taking in Africa and Asia, continents that still had large unclaimed and untamed regions. The United States, on its own continent, expanded significantly to its final western oceanic boundaries. The American Civil War was the world's first attempt to settle the question of African slavery (I say the world's first attempt because no other country at the time thought of itself as having slavery, though in years to come the class structures in Asia found themselves illegitimatized in comparison to America's post civil war equality). There were other wars: The War of 1812, the Boer War, the Boxer Rebellion, and the Crimean War. Alfred Nobel invented dynamite and ushered in the era of "the bomb." Mechanized war, however, waited for the 20th Century and its world wars.

At the end of the 19th Century, the American labor force[2] was 73% male and comprised only 34% of the population. The U.S. population of 1899 numbered 67 million compared to over 281 million in the year 2000. The dominant employment for workers aged ten and above was farming, totaling eight million. Domestic services, servants and other related labor, were next, with around four million. Then in decreasing order were bookkeepers, merchants, carpenters, dressmakers, steam railroad workers, teachers, launderers, book makers, blacksmiths, cotton mill operators, machinists, iron workers, restaurant labor, butchers, doctors, lawyers, clergy, bakers and finally the military with thirty thousand soldiers.

In 2000 the distribution of labor was startlingly different from that at the end of the 19th Century: 46% of the population worked instead of 34%, the minimum work age increased from ten to sixteen. Women composed nearly 50% of the workforce. In 2000 the dominant employment group was made up of those in management and professional occupations with 34% of the total workforce or 44 million people, in which men and women were employed equally! Farming, the 19th Century's biggest employer, was down to a paltry two million with only one million actually actively engaged in farming; the remainder were managing, distributing and selling what was produced.[3]

This single change, from farming to the professions, makes plain the magnitude of the social change that occurred over the span of the 20th Century. The number of college and high school educated people necessary to effect this change is also illustrative. A high school education was achieved by 80% of 25 year-olds by the year 2000, of whom 23% later completed a bachelor's degree or better. In comparison, the 1940 census (being the first such census data available) showed that only 25% of the population had a high school education and less than 5% had completed a college degree.

The second largest occupational group after the professions was the sales force in the year 2000, including those in offices supporting sales, at 35 million or 27% of those employed. The remaining three occupations were:

1. Production and transportation of goods at 19 million;
2. Service workers at 19 million including 6 million food preparation workers, 4 million building cleaners and 2 million in protection services such as fire and police;
3. Construction and maintenance workers with 12 million employed.

In 1910, 66 million people lived in rural areas and only 26 million lived in metropolitan areas. By the year 2000 fewer people, 55 million, lived in rural areas, but by then a whopping 226 million people lived in metropolitan areas. City center growth over that span of time would be small, but suburban growth around the cities was enormous.

In 1910 the suburbs held 7 million of the 26 million metropolitan population. In 2000 the suburbs rose to 50 million, a sevenfold increase, whereas city growth increased from 21 million to only 30 million over the same time span. This city suburb concentration did, as you will see, change the innovation rate and location of innovation in the 20th Century.

· · · · ·

You wake today in 1899 to the sun rising: it is winter, horses snort steam from their nostrils and whinny, heading toward the city. You smell coal as your central heating stove recovers from the night's cooling into full heat. Wood shoved into the cooking stove will bring you fried eggs this morning since it is Saturday. Your mother will be hand sewing your dress for school on Monday and you will walk those miles in snow, hoping and praying for the day when your family will own the vaunted automobile. You wonder what else is coming with all the changes. You have heard of those who cook with gas stoves. You have seen some automobiles owned by the rich. You can travel to your local city on the steam train if you are willing to pay the fare. Tonight you and your friends will meet late at a local shop to have sodas and talk in ways your mom and dad would not approve. Tomorrow you must play the piano at the church gathering for the afternoon youth group and you have not yet mastered the music, but you will. The sun is welcome in these cold months where vegetables are scarce and potatoes and carrots dwindle. Thank goodness for the chickens in the yard!

Reference notes

1. Web survey of graduate universities worldwide as of 2004: Americas—2759, Europe—1600, Asia—1251, Russia—900, India and Pakistan—800, China—580, Japan—484, Africa—350, Middle East—350. (UNESCO—International Association of Universities, www.unesco.org/iau).
2. Based on 1890 US census data.
3. Combination of 1890 Special Census Report on Occupations and Occupations 2000, Census 2000 Brief.

CHAPTER 1

Table 1 Social Change

CHANGE EVENTS	YEAR	CHANGE AGENT
Women's Right to Vote	1906	Finland's Diet (parliament)
China Abolishes Slavery	1910	China
First American Birth Control Clinic	1915	Margaret Sanger
Bolshevik Revolt	1917	Vladimir Lenin
First Congresswoman	1917	Jeannette Rankin
Prohibition — the Volstead Act and the 18th Amendment	1920	WCTU/Anti-Saloon League
Progressive Civil Disobedience	1920	Mohandas K. Gandhi
Music and Medicine for the Poor	1920	Albert Schweitzer
19th Amendment — Women's Vote	1920	C. Stanton/ S. B. Anthony
Modernized Turkey for Muslim Women	1923	Mustafa Kemal Pasha Ataturk
Sports Hero Model	1927	Babe Ruth
Stock Market Crash — American Depression	1929	Wall Street USA
Major Construction Projects — Hoover Dam	1933	Stephen Bechtel
Entitlement, Welfare, The New Deal	1933	Franklin D. Roosevelt
Prohibition Repealed (21st Amendment)	1933	USA people
Alcoholics Anonymous (AA)	1935	Bill Wilson
The General Theory of Employment, Interest and Money Deficit Economics	1936	John Mayard Keynes
Capitalism/Organized Labor	1936	Walter Reuther
Commonsense Baby and Child Care	1946	Benjamin Spock
African American Sports	1947	Jackie Robinson
Universal Declaration of Human Rights	1948	Eleanor Roosevelt
Male Sexual Behavior	1948	Alfred Kinsey
Desegregation, One Man-One Vote	1953	Chief Justice Earl Warren
McDonalds — Father of Fast Food	1955	Ray Kroc
Beginning of the End to Racism	1955	Rosa Parks
Marxism on Chinese Social Order	1958	Mao Zedong
Oral Contraceptive (1923, C. Djerassi)	1960	Pincus/Min-Chueh Chang/ Rock
Power to Overcome Incapacity	1960	Helen Keller
OPEC Oil Price Control	1960	OPEC

Table 1   Social Change

CHANGE EVENTS	YEAR	CHANGE AGENT
Peace Corps of America	1961	John F. Kennedy
Folk Music Re-Invented	1963	Bob Dylan
Feminine Mystique	1963	Betty Frieden
LSD-Chemically Altered Perception	1963	Timothy Leary
Economic Philosophy — The "Knowledge Worker"	1964	Peter Drucker
Black Power Awareness	1964	Malcom X
Miniskirt	1965	Mary Quant
Social Science-fiction TV Philosopher	1966	Gene Roddenberry
Hippies Summer of Love	1967	Haight Ashbury
Revisions to Spirituality	1968	Eastern Religions
Powerful Woman Politician	1969	Golda Meir
Feminism	1970	Germain Greer
Telethon Fund Raising for the Sick	1970	Jerry Lewis
Economics — Government Monetary Policy	1970	Paul A. Samuelson
Roe vs. Wade — Abortion Rights	1973	US Supreme Court
Future Revealing Movies	1977	George Lucas
Dedication to the Helpless	1979	Mother Theresa
Government Regulation of Economy	1979	Paul Volker
First Female Supreme Court Justice	1981	Sandra Day O'Connor
Future Revealing Movies	1982	Stephen Spielberg
HIV and AIDS Epidemic Scare	1983	Gene Mutation
Revisions to Spirituality	1988	Babba Ram Daas-R. Alpert
End of Marxism	1989	Mikhail Gorbachev
Muslim Re-Integration	1989	Salmon Rushdie
Financial Failure of Marxism	1989	USSR
Self Help Book Mania	1990	Baby Boomers
Integration Begins in S. Africa	1994	Nelson Mandela
Over-Population	1999	India

"You will find, throughout this book, that the most visible names attached to an event or discovery are often only pointers to the ground swell of contributors who participated . . ".

SOCIAL CHANGE

THE LARGE CHANGE IN SOCIAL BEHAVIOR

The changes to society during the 20th century have been phenomenal by comparison to those that crept in over the threshold of the preceding half-millennium. To be sure, the basics of home life in the modern industrial world were present in ancient Rome, but much has changed.

Many of the social changes noted here had their epicenters spread across the Americas, Oceania, the USSR, Europe and Japan. Outside those regions, social changes were beginning to reach China, India, and North Africa by mid century, but had left the largest portions of mainland Africa, rural India and Southeast Asia untouched. [1]

As an example, the enfranchisement of women became law first and last in Europe (Scandinavian countries included), and most broadly in the United States by 1920. The list of active women worldwide, those working on issues important to their equality at the turn of the century, is long and their names more symbolic than complete. You will find throughout this book that the most visible names attached to an event or discovery often point to the groundswell of contributors who participated, and those women listed must be viewed as symbols more than absolute truth regarding who was first and deserves most credit.

The social changes that affected multiple millions of people during the 20th Century boil down to these three summarizing characteristics of modern society:

- to vote and work as equals, unaffected by race or gender;
- the use of deficit economics in government, the emergence of the knowledge-worker and fairer labor-management interactions, the dwindling appeal of pure Communism and the emergence of worldwide economic democra-preneurship;
- the relaxation of sexual mores and the failure of the Prohibition movement, which exposed weaknesses in the use of democratic-moralism.

Suffragists' march on Washington, 1913

TO VOTE AND WORK AS EQUALS.

During the 20th Century the protections of the Bill of Rights in the United States were expanded to include government guarantees of universal suffrage and equal access to work, these guarantees to be independent of race or gender. This extension of governmentally guaranteed rights expanded across the world throughout the century, beginning with providing women the right to vote in the early 1900s, and continuing with the injection of large numbers of women and African American workers into the economic system — for the first time as equals. Though women and subjugated races had been part of workforces in many countries for centuries, the idea of equality with white males in the United States, for example, was a major change in social uniformity. This change is not yet worldwide, but is clearly becoming so in this 21st Century.

Hotel Clark, 1939

The women's movement and the efforts to eliminate racial bias had their roots in the late 1800s starting with the abolitionist movement worldwide and the Civil War of 1865 in the United States. One can easily find a list of more than thirty women active in the leadership of an equal right to vote, with or without ownership of property, in the early decades of the 20th Century. Real voting access for the Negro, or as he was later called the African American, was delayed at least another forty painful years.

The 20th Century struggled with two other human rights of a very different nature; the right to health-care and the right for equal access to work and equal access to the conveniences of life-support stores (hotels, grocery stores, department stores and public transportation) for the disadvantaged.

The worldwide discussion of state subsidized access to health-care was more an evaluation of an evolving program than a human rights breakthrough of the 20th Century. Before the discovery of antibiotics that cure bacterial infections, there was little doctors could do to cure the more serious illnesses. Increased understanding of the human body via the biological and chemical sciences during the 19th and 20th Centuries was responsible for society's evolution toward the assumption of equal access to national health care. By the end of the century, the assumption that everyone deserved access to health care and the dilemma that not everyone could pay for such protection via insurance became the defining contradiction of health-care. At present those who can't afford insurance go uninsured, but society provides emergency care in spite of an individual's inability to pay. As a result, emergency rooms are going bankrupt. The 20th Century created this social change and its contradiction, but did not resolve it. The costs are very likely to force some sort of resolution in the 21st Century.

A fourth and last, but still nascent portion of the extension of government managed individual rights — entitlement of the disabled — gained support across the United States and Europe in the latter part of the century. The concept is not nurtured in most other countries or cultures. For the United States and Europe, work places are required to accommodate disabilities that do not interfere with the work required. As medicine improves, the number of disabled is likely to stay low enough that the social cost and business cost of this "right" will not become impossible to support, unlike the problem of cost that universal access to health care creates.

NEW ECONOMICS, EMERGENCE OF THE KNOWLEDGE WORKER, FAIRER LABOR PRACTICES, DWINDLING APPEAL OF PURE COMMUNISM, WORLDWIDE POLITICAL ECONOMIC DEMOCRA-PRENEURSHIP[2]

The 20th Century's greater fairness and equality in access to vote and to work significantly strengthened the capitalist formula and reduced the appeal of the philosophies of Karl Marx, as implemented in 20th Century Communism. Also the achievements of unions in 1930's America restrained the sometimes heavy hand of management, further improving working conditions and wages. Franklin Delano Roosevelt's New Deal provided a solid platform from which labor negotiated contracts with management. Eventually FDR managed to understand and use Keynes' proposed deficit

Women welders, Erie, PA dry dock, 1943

spending approach[3] to pull the United States out of the deepest (in fact worldwide) financial depression in modern times. This change to using deficit spending to finance government wellbeing continues today and protects the vigor and resilience of the 20th Century's evolving version of Capitalism. Still later, minimum wage laws and judicial rulings based on the uniform application of the law (as embedded in the 14th Amendment) reversed entrenched, state-controlled inequities in wages and discrimination based on gender, race and creed.

Synchronous with these United States' mandated changes of fair treatment was the increasing educational depth of the populace. Better college education came about through the use of state-funded, low cost state colleges (See the statistics given in introductory chapter on 1899). In addition, the GI bill paid for college educations for military personnel after World War II. Consequently, a new manner of worker emerged, best described by Peter Drucker as the "knowledge worker."[4]

"Knowledge workers, even though only a large minority of the work force, already (1994) give the emerging knowledge society its character, its leadership, its central challenges and its social profile. They may not be the ruling class of the knowledge society, but they already are its leading class. In their characteristics, their social positions, their values and their expectations, they differ fundamentally from any group in history that has ever occupied the leading, let alone the dominant position. In the first place, the knowledge worker gains access to work, job and social position through formal education."

Drucker's term, the "knowledge worker," described these college trained technical specialists who were no longer "apprenticed" and therefore shackled to their specialty. Instead they came with newly portable skills to businesses in which they were capable of working from the first day. Their emergence signaled a change in worker treatment and mobility. As their skills matured in the work place they continued to retain portability, increasing the value of their maturing knowledge. Their portability of knowledge was true for engineers, chemists, biologists, journalists, accountants, sales and marketing personnel — the entire gamut of the college-educated individual.[5] As a consequence, employers sought these highly valued employees and so upgraded working conditions to encourage their retention.

The revolutionary ideas from the beginning of the 20th century that fomented national revolutions in China, Cuba and Russia seem to have dissipated over the century; or more explicitly, merged into one economic river in which we are all immersed. The increases in mobility and communication provide easy access to goods and employees across the globe. In the end, the fall of the Berlin Wall in 1989 did not signal, as many have suggested, the end of Communism so much as a reduction in the importance of differences between Communism and Capitalism — the two dominant economic philosophies of the 20th Century. In the United States and European democratic countries, a social middle-ground was formed in which electoral processes determined economic and social structure without revolution. The world's economic river is now more tied to governmental policy and its financial controls, corporate product successes and worldwide economic health, than to the form of government itself. We have become aware that an economic depression in one major consuming or producing country causes economic consequences in many others. World interdependence has become much more visible to every human being.

DEMOCRATIC-MORALISM, BIRTH CONTROL PILLS AND THE RELAXATION OF SEXUAL MORES

The year 1920 ushered in a popular constitutional amendment, the policy of Prohibition, which made the sale of alcoholic beverages illegal in the United States. This was one of a continuing series of efforts by self-declared and hopeful moralists seeking ways to control human over-indulgence or, as some would have it religiously, sin. Their attempt to outlaw alcohol combined with the equally demanding forces upon those who wished to purchase it led to extensive illegal distribution and lawlessness. More people were breaking the law than were upholding it. This social experiment in morality by democracy came to an abrupt end a tawdry thirteen years later, when Prohibition was overturned by a majority vote amending the Constitution of the United States by revoking the original amendment. Prohibition should serve as a signpost to help us avoid future embarrassment with legislated morality.

By late in the century the United States had been through several attempts at legislative control of human indulgence and excess. Success at this law-driven morality was limited. The addition of mind-altering and addictive drugs to the moral control battlefield only exacerbated legislation and its policing. Similar moral legislative battles emerged in attempts to define lawful abortion. Those battles continue. There have been some, though few, vigilante acts of violence against the present abortion law, but none as destructive or violent as prohibition's gun battles. Society is still struggling with unwanted pregnancy control options, most recently pressing the U.S. Supreme Court for guidance.

SLOW EVOLUTION OF CULTURAL MORALITY

Cultural morals change very slowly. Our present sexual mores hark back hundreds of years. The author's limited single-generation observation suggests that moral standards tend to take more than three and probably five generations to change substantially. One hundred and twenty years or more! The 20th Century may have sown the seeds for moral change but likely the 21st will not see their result.

Polarized extremes of moral advocacy split society and do not seem to have solved the problem being addressed. This bimodal social distribution increases the potential for violent conflict. The 20th Century's post-prohibition alternative for the control of the abuse of alcohol was to use social pressure to stigmatize excess, assisted by curative self-help organizations such as Alcoholics Anonymous and its twelve-step program. This manner of discouraging the worst excesses works to reduce the actual problem without the inherent split to society, and without the violence. Recent legislation to arrest drivers who test over a certain percentage of blood alcohol has also implemented a quantitative middle ground of morality that seems to be helping reduce drunk driving and its deaths.

Change in cultural morals occurs only when widely altered physical and emotional realities pressure old rules toward change. There were three events that strongly pressured sexual mores to change during the 20th Century in the United States and Europe.

First, the introduction of the birth control pill significantly altered the consequences of premarital and extramarital intercourse. The birth of outside-of-marriage children in earlier centuries created such extreme public outcry and social dislocation that intimacy with "banned" partners was strongly sanctioned and social penalties were severe — to and including death to the participants. The Pill significantly reduced the risk of these undesired children.

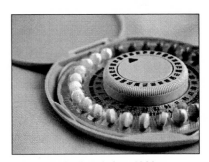
Birth control pill pack, late 1900s

Second, the availability of voluntary sterilization by minor surgery further reduced the risk of unexpected pregnancies.

And third, the 1973 Supreme Court ruling in Roe vs. Wade extended 14th amendment rights to protect a woman's private use of abortion to terminate an unwanted pregnancy during its early stages.

The women's rights movement early in the century contributed to birth control advocacy, helping pill developers envision and predict an emerging market. In 1916 Margaret Sanger was a vocal advocate of birth control as a woman's means of controlling her body and her life.

Later Alfred C. Kinsey presented evidence in his reports issued in 1948 and 1958 of these changes in attitudes about sexuality that were forcing change on our cultural mores. In addition to Kinsey, the hippie free-love era of the '60s also placed the newly gained sexual freedoms before the world for review. Though much of that time's openness has disappeared or gone underground, it presaged the opening of many questions and the consequential relaxation of behaviors regarding intimacy outside of marriage. Those questions may not be answered or new mores established by the end of this 21st Century, but change is taking place. Resistance to change is also evidence of change. The Roe vs. Wade decision, and its subsequent choices, polarized society into two moral camps for at least the next fifty years; the question being, "When is abortion murder? "

The sad consequence of polarization seems to be the reduction in public effort to solve the problems of unwanted childbirth: abuse of unwanted children, the birth stigma to unmarried mothers, and the sad deaths of embryos. These all have been neglected in favor of legislative attempts to prevent abortion. More middle-ground laws must be sought in an effort to prevent this pitting of large populations, albeit minorities, against one another. Portions of society must become more moderate in their attempts to limit human differences and more accepting of the full range of human variety for the best solutions to be found.

Reference notes
1. Population Reference Bureau *2004 Population Data Sheet*
   http://www.prb.org/wpds
2. Defined as democratic government combined with entrepreneur driven industry.
3. John Maynard Keynes, 1936, *The General Theory of Employment, Interest and Money.*
4. The term Knowledge Worker was first used by Peter Drucker in his book *Managing for Results,* 1964 "Even the small business today consists increasingly of people who

apply knowledge rather than manual skill and muscle to work." And later employed in his book, *Post-Capitalist Society*, 1993. "Instead of capitalists and proletarians, the classes of the post-capitalist society are knowledge workers and service workers." And finally in the Edwin L. Godkin Lecture at Harvard's Kennedy school; *Knowledge Work and Knowledge Society The Social Transformations of this Century*, Peter F. Drucker May 4, 1994

5. *The Wall Street Journal*, December 26, 2003, Robert Reich, Labor Secretary for President Bill Clinton, coined the term "Symbol Analyst" for the same work function as the "knowledge worker" and described that job vs. the in-person service worker as the two expanding job arenas in the United States separated by diverging hourly wages, one increasing and the other static or worse.

CHAPTER 2

Table 2   Mobility and Communications

CHANGE EVENTS	YEAR	CHANGE AGENT
Gyroscope (MIT)	1901	Charles Draper
Transatlantic Wireless Telegraph	1901	Guglielmo Marconi/Karl F. Braun
Powered Flight	1903	Wilbur and Orville Wright
Offset Printing becomes real	1904	Ira Rubel
Vacuum Tube Rectifier — Detector	1904	Sir John Ambrose Fleming
Automobile Assembly Line	1908	Henry Ford
RF Amplifier Vacuum Tube, Triode	1908	Lee de Forest
Electro-Magnetic Waves	1910	P. Zeeman/ H. A. Lorentz
Oil Cracking for Gasoline	1912	William Burton
Super Regenerative Radio Amplifier	1914	Edwin Howard Armstong
Sound Track Movies	1922	T.W. Case
TV Camera Tube (orthicon, vidicon)	1923	Vladimir Kosma Zworykin
The Loud Speaker	1924	Rich Kellog
Rocket "for Extreme Altitudes" (V2= 1944)	1926	Robert H. Goddard
Negative Feedback Amplifier	1927	Harold Black
Quartz Clock-accurate time keeping	1927	W.A. Marrison & J.W. Horton
Real Sound Movies *The Jazz Singer*	1927	Warner Brothers
Revolutionizes Road Transportation	1929	Autobahn — Germany
Magnetic Wire Recording (V.Poulsen)	1928	J. Begun/ F. Pfleumer
For Whom The Bell Tolls — We are one nation	1929	Ernest Hemingway
First Electrical Computer	1930	Vannevar Bush
Gallup Polls — Reliable Polling	1935	George Gallup
Airlines — Pan American	1935	John Trip
Helicopters — Vertical Lift	1939	Igor Sikorsky
Cathode Ray Tube (CRT)	1939	Zworykin/Farnsworth/Dumont
Magnetic Tape Recording Heads	1941	Marvin Camras
First Jet Engine Aircraft	1941	Sir Frank Whittle
Spread Spectrum Idea, later used for cell phones	1942	Hedy Lamarr & George Antheil
Mathematical Encryption/Security	1945	Enigma, Bletchley Park
The Transistor	1948	Brattain, Bardeen, Schockley
Cashless Society — Credit Card	1950	Diner's Club

Table 2   Mobility and Communications

CHANGE EVENTS	YEAR	CHANGE AGENT
Zone Refining of Silicon	1952	William Pfann
Color TV	1953	RCA-NTSC & Shadow Mask
Rock and Soul	1955	C. E. Anderson "Chuck" Berry
Fiber Optics	1955	N Kapany/B O'Brien
Same Day Global Interconnection	1956	Airfreight
Video Tape	1956	Charles Ginsberg/Ray Dolby
Container Ship — Ideal-X	1956	Malcom Mclean
Xerox Dry Copying	1958	Chester Carlson
Laser/Maser	1958	Gould, Townes & Schawlow
Integrated Circuit	1959	Robert Noyce/Jack Kilby
Ruby Laser	1960	Theodore Maiman Bloembergen
First Manned Satellite	1961	USSR, Gagarin
1st Communications Satellite Telstar I	1962	AT&T/NASA/Bell Labs
Laser and Light Emitting Diode (LED)	1962	Robert Hall/ Nick Holonyak Jr.
Rock and Roll world wide	1963	The Beatles
Fiber Optic Cable/Communication	1968	Corning Corp
Fiber Optic Telephony (wide use 1977)	1970	AT&T
Public Keys, Encryption	1970	Martin Hellman/M Williamson
RSA Public Key — Diffie-Hellman	1976	Whitfield Diffie/Martin Hellman
Cell Phone	1977	Cooper at Motorola & AT&T
Apple II	1977	Wozniak, Jobs
Cheap High Quality CD Music Reproduction	1979	Joop Sinjou & Toshitada Doi
Fast Mail	1985	Federal Express Corporation
Global Positioning System	1987	J. J. Spilkher
World Wide Web (WWW)	1989	DARPA, UCLA, CERN, MIT- Mosaic, HTTP, HTML, TCP/IP, DNS
Easy Graphical Transmission	1990	FAX

Satellite photo shows massive highway intersection near San Francisco

"As humans have done through each major stage in social development, we will again absorb these new tools . . . into the mainstream of life, invent a revised social order, and soon take this sped-up mobility for granted."

Mobility
and Communications

The 20th Century is likely to become most widely known for the technologies described in this chapter, though their society-wide impact can best be characterized by the presence, at century's end, of the five aspects of modern life summarized below:

1. The availability of low-cost, worldwide communication and access to information, both open and private (encrypted), from your living-room or office;
2. Worldwide, low cost availability of images, movies and music that provide a universal language through which to share emotion and experience;
3. Fast, closed-loop society-wide feedback systems such as polls and votes;
4. Planetwide transport that provides low cost access to goods,

information and people over short and long distances;

5. A paradigm change in the way technological advances are made; rather than the few super-scientists of earlier centuries, thousands of scientists, engineers, business men and women now work together to make the many small advances that, only in concert, change the world.

The first four items refer to effects of technological advances, while the fifth item proposes a new causal relationship.

Because this chapter reviews aspects of the 20th Century most frequently written about in the press and literature, this book will not cover in detail any descriptions of gadgets, conveniences, or stories of lives saved or taken to further embellish the electronic era.

What this chapter will cover is a list most of us have not heard of that describes the inside of the mobility and

Apple's 1984 Macintosh (SE model, first with hard drive in 1987)

communication revolution. You should spend some time looking over the list itself, maybe even browsing the web for answers to explanations of some of the items that seem inscrutable without the Wikipedia's translation into graspable definitions.

This next example, derived from the list, is an unheard-of detail absolutely necessary to modern electronics. It begins to illustrate the fifth summarizing aspect of major changes in living during the 20th Century; that the major names in science are not the ones who made it all happen, but instead, the unheralded engineer inside the gears of the technology machine.

THE ZONE-REFINING OF SILICON

The mastering of two very fundamental technologies was necessary before the explosion of transistor and silicon manufacturing could take place for hand-held electronics. The transistor, an invention fundamental to small electronics, had been imagined fifty years prior to its actual invention in 1948; but it had not been built or verified until these two technological problems had been solved in the lab at Bell Labs in 1948. The first problem was to obtain a clean silicon surface, molecularly clean, without contamination or oxidation. The second was to create mono-crystalline silicon in thin, large diameter disks. In 1952, shortly after the invention and verification of a single laboratory transistor, zone refining was invented. This discovery was more an implementation than invention. The process used an amorphous crystalline blob of silicon (beach-sand's main constituent) and moved it vertically through a round coil emitting high power radio frequency energy which melted the long blob (or ingot) of silicon through its cross-section. This is similar to microwave heating a chicken pot pie, and, as it passes through the coil and cools, it solidifies as a single crystal structure ready for slicing with a diamond saw into the thin, oxide-free disks needed to create the very small electronics of amplifiers by using photo lithography and chemical etching. All these technologies were implemented in the 20th Century[1].

Using this example as a beginning we will take the five aspects summarized at the start of this chapter, one by one, allows us to touch on many of these less known details in context from this chapter's list of events, inventors and inventions.

FAST CLOSED LOOP FEEDBACK

Previous centuries were characterized by phenomena occurring locally and the effects of those local changes spreading slowly across many borders and geographic boundaries, eventually reaching the limits of civilization. Before the availability of aircraft, electronics, copper and optical cables, the movement of information and ideas generally took months to reach a new destination, months to be disseminated to the human community in those far-flung places, and months and even years to generate a common response that then *might* return to its origins. This round-trip feedback loop, and its former year-long time span, changed dramatically in the 20th Century.

In addition, the development of mathematical, machine-assisted encryption provided the equivalent of house key privacy in communicating and exchanging this information. Money transfer and management of stock investments are now essentially done by remote control using encrypted communications over thousands of miles. I sit in my living room easy chair activating multiple thousand dollar transactions in locations that I am in fact unaware of and potentially as far away as Zanzibar.

The speedier closing of the loop improves people's and organizations' ability to either correct errors or to change their plan. We now see market forces respond within minutes to lending rate changes and to unemployment figures that signal a change in economic health within weeks. The effect of a competitive new company becomes visible to the market within a business quarter. Of course there is haste in the conveyance of news regarding scandal, war and natural disasters. The fast closing of the societal feedback loop can make society more stable in the same way it does an electronic system. And as in electronics, greater stability is true only if the feedback is used to turn the right societal knobs, making changes with the right timing for the machine of society to settle down. Of course feedback can make governments and organizations change too rapidly as well, causing erratic behavior, where there once was slow change. Delays in assessment previously led to misreading cause and effect. Now misreading is no longer necessary, but humorously is still an option. Corrective efforts can now be made and their effect observed as the change's effect takes place, often called the rate and direction of change.

LOW-COST ACCESS TO COMMUNICATION AND INFORMATION

Around the beginning of the 20th Century, Marconi and Braun separately discovered the groundwork sciences for wireless communication and demonstrated them to an awestruck world. Marconi is only retrospectively the most visible scientist of radio.

Many others were doing the groundwork science and engineering for communicating without wires at the same time as Marconi's and Braun's work was publicly moving forward (wired telegraph Morse code had routinely been in use at the end of the 19th Century). Marconi most dramatically demonstrated continent-to-continent telegraphy to the world in 1901 by sending transmissions of intelligence — intelligence as elemental as buzzing in headphones — through the air to far-flung receivers where someone heard those same signals. The evolution of our modern communications began with the Marconi and Braun discovery that an electric arc, similar to that of a spark plug, could produce an effect (a signal) that would move through the air and be heard at a distance by

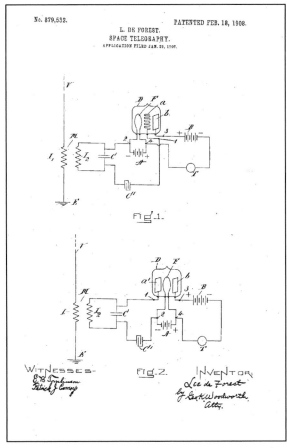

Lee de Forest's drawing of a triode in his 1908 patent application

using a glass tube containing iron filings to sense the electrical disturbance sent over that distance by the spark.

By 1908 de Forest's triode valve or tube was invented and thus, we had a crude and short-lived receiver and amplifier, a device critical to the receiving of very small signals from a remote location and to making them stronger than the electrical noise surrounding us in the air at all times.

In spite of the triode's invention in 1908, the first wide use of radio took more than a decade to come into its own in the public arena. That decade consumed hundreds of engineers in sometimes parallel efforts to refine the functioning of the radio receiver.

Let us take a short side trip to dig into the construction and function of the triode and its ability to amplify an incoming intelligent signal which provides the listener with sound reconstructed from a far away radio station.

Amplifier is a common enough word now, but it was not then. The triode is a glass bottle with the air removed (to prevent the metal inside from burning up) and three pieces of metal inside (see the previous page patent drawing portion labelled Figure 1). These metal objects form the three ports or access points to the triode; one conveys the signal into the device, one makes the signal available to the next stage of amplification after exiting the triode and the third is the electrical return connection for both — the ground connection, as it were. If you remember the Electric Company, a derivitive of Sesame Street, you may remember that a circuit means circle, the electrons must move through a complete path to make a connection. Each input wire has to have a returning wire to complete its circuit. How do these three connections amplify?

A very small signal collected from the air by a wire antenna and containing the information we want to hear goes into the *grid*, or the in-port, of the triode (*a*). It is a signal that is one thousandth of the strength needed for the human ear to hear. This weak signal from the air (the ether) is amplified as a result of its presence on the *grid*, thus controlling the flow of the electrons going from the common point, the cathode (*F*), to the output port, the anode (*b*). To a large extent, the in-port, or *grid*, acts as a gatekeeper or dam of the electron river that flows from cathode to anode. The *grid* acts like the knob on a water valve, where a small twist of the knob lets a large amount of water through compared to the force applied to the knob.

The reason the water valve amplifies flow-volume in a pipe-valve is the mechanical advantage provided by the screw thread in the knob. For the triode amplifier, the reason for reduced or increased flow in the electron stream that moves from cathode to anode is that the *grid* is physically in the way of or barring the way of electrons that want to flow from cathode to anode. The grid is placed at the best location to repel the electrons trying to flow from one end of the triode

1940's Tube Radio

to the other while consuming the least power. This change from a low power signal controlling a larger power signal is amplification.

Fifty years later, amplification was provided by a device one-hundredth the size of the triode, the solid-state transistor. The transistor is again a three-port device wherein by wiggling the input, now called the *base* instead of the *grid*, a small signal allows a larger copy of itself to emerge at the other port, the *collector*. Further strengthening occurs after a string of these amplifiers makes the signal strong enough to blast forth from your stereo. Or, alternately the amplifier will make a weak television signal that would otherwise appear as snow strong enough to form the clean, clear picture of an Olympic ice skater on your television screen. Thank you 1908 triode.

I say triode instead of mentioning de Forest because many people were pursuing a triode-like result and it is likely the person's name is less important than the group of people in pursuit and the triode's eventual accomplishment.

Thereafter each of many hundreds of inventions, feats, and steps occurred that pulled us further and further into speed of light communication, which culminated in both satellite and fiber optic cable-enabled worldwide electronic connections, and cell phone enabled mobile communication.

IMAGES AND MUSIC

The development of visual communication unfolded in parallel to the connection of people by voice. This was assisted by the evolution of the radio-like, tube-based television receiver with the added invention of the CRT (cathode ray tube). Beginning with film and movies in the '20s and black-and-white TV in the late '30s, color TV and video tape in the '50s and the personal computer in the '80s, worldwide communication through images truly came of age. The TV phone had to wait for the 21st Century in spite of futurists' promises at the 1950 World's Fair. By 1995 most necessary information was being transferred by voice and image in real-time, at costs that were within reach of anyone working at average personal income levels in the industrial nations.

In the 1960s, the solid-state gallium-arsenide laser and the LED (light emitting diode) made it possible for a recently developed metallization on a plastic disk, the CD and DVD, to store and replay high quality movies and music. Again this used an amplifier on the signal picked up when a light source (laser) was shined on the disk and its reflected light variations directed to electronics designed for that purpose.

The integrated circuit, the IC and its digital cohort — the microprocessor, provided amazing electronic functionality in constantly smaller packages, culminating late in the century in the iPOD™ and the cell phone the size of a cigarette pack. Some engineers would suggest that the IC and microprocessor are even stronger examples of the immensely powerful consequences of physicists, engineers and technology joining up to change the world. There were thousands of inventions, tens of thousands of papers, and millions of dollars and employees involved in the miracle of recording and reproducing images, movies and music for us. The mass distribution of transistorized radios and Hi-Fi electronics (dependent on the invention of the transistor and electro-magnetic speaker) were also critical in bringing the world this universal language of music.

INTERNET

In 1989 the Internet arrived, originally called the ARPANET (Advanced Research Projects Agency Network, in 1969). It was a means of sharing inter-university information allowing for the fast dissemination of scientific and militarily relevant advances. Later the emergence of small, intelligent, information-processing machines (computers) in each home connected by the telephone system, gave each owner of a telephone access to all information connected to other telephones in far-flung universities, and eventually other homes and other huge reservoirs (disk arrays tied to web servers) of information. This access, given at low cost to everyone, has changed the world's ability to restrict information to geographic areas.

PLANET TRANSPORT

Other 20th Century instruments of mobility and communication not yet discussed, but obvious from the list are: the commercial jet airline, oil tankers and container ships, and the automobile combined with Eisenhower's highway system in the United States and the Autobahn

Container ship being loaded from trucks at lower right

or Autostrada in Europe. All of these provided people access to almost anywhere on the globe and enabled the shipment of food, manufactured goods and oil to all nations. The rise of the commercial airline and the trucking container industry caused the previously shining star of the railroads to decline in the United States during the latter part of the century, though trains' importance could resurge in later centuries.

The emergence of the container business, wherein railcar sized metal boxes are loaded with a product at the manufacturers location (often in Asia these days of low labor cost), trucked to a shipping port and loaded onto large container ships, has lowered the burden of shipping cost enough to allow worldwide manufacturers to compete without significant cost disadvantages. For example, a bottle of Australian wine shipped to Europe will cost only 12 Euro cents more than wine made at the same labor cost in France. Every land and sea transport vehicle has been redesigned to

accommodate these container boxes, first used in 1956. The ships' holds are designed to enable speedy loading and unloading as well as safe and economic ocean transport of up

1935 DC-3, the workhorse of piston-driven airliners, manufactured by Douglas Aircraft, owned by Howard Hughes.

to six thousand such containers. Sixty percent of all goods shipped are housed in these standardized containers. Rail and truck transportation companies compete for container transport though almost all local delivery is by truck.

The worldwide access to such transportation as international flights that cost only two months' wages, or automobiles costing closer to a year's salary, enables unprecedented distance to be covered at will over the span of one's lifetime. Goods are exchanged from all corners of the world, providing low bottom lines as a result of competition based on the cost of labor. These low-cost countries' economies gain preferred access as competitors to the consuming world economies. Individuals, as well, gain access to all parts of the world for personal pleasure and access to knowledge. One can view the significant elements of cultural history and intellectual achievement for Western and Eastern civilization nearly at will, gaining immense retrospective insights. This material and personal worldwide mobility better integrates our lives as a community of residents on the planet.

BANDWIDTH = COMMUNICATION

The key image and data communication table-of-elements, or critical components as it were, is best summarized as follows:

For good communication to exist, one requires a means of 1) transmission and reception, 2) connection, 3) storage, 4) replay and display. In some ways human air travel (as well as container ship transport of goods) can be broken down in the same way; the plane and air form the transmission and reception media, the customs agents, harbors, airports provide the connection means and we, ourselves, are the storage, replay and display vehicle. (In the case of products, storage is in the form of warehouses and distribution centers and use again is in the hands of the human receiver; the equivalent of replay and display for information.) Our mind and eyes record what we see and hear for later recollection. Movies however, have a different replay and display process by projecting images onto the screen in specially designed arenas, bringing recorded media of the past back to life. All of these 20th Century media and shipping methods have primary limitations in the quantity of real-time information and goods that can be delivered per unit of time, often called "bandwidth."

It is this final commodity, bandwidth, which, like water and energy, will become the way we characterize a given era in mobility and communication's growth. When the per-capita amount of bandwidth flattens and reaches a maximum, this alone will indicate the maturity of communication as a technology. That flattening for electronic information bandwidth was far from being reached during the 20th Century. The container and oil shipping industry also seemed far from flat per-capita growth in the century.

Clearly this century has altered human connection. You can essentially *live* anywhere you wish instantly, via electronic communication and visualization, without physically moving from your home in Kansas or Singapore. The only invention that remains to be created, as imagined in the science fiction stories of *Star Trek* and Isaac Asimov, is a means for instantaneous material transfer as in "Beam me up Scotty," or a means of time travel, best described in Michael Crichton's novel, *Timeline*, and the 2004 movie, *Primer*.

The most society-changing consequences of such worldwide community access to information were: first, — to remove government and corporate control of the distribution of information and second, — to provide a capacity to quickly distribute information to many people concerning proposals and actions, and to obtain instant feedback from them. These changes reduce the possibility for abuse of information by corporations and governments. Before 1989 and the fall of the Berlin Wall, the USSR did not allow its citizens to read the *New York Times*, available in their libraries, without explaining what they wanted to read *apriori*! Today one can read any newspaper one can translate by using the Internet access to the World Wide Web. Abuse now more frequently finds its place within marketing's control of content in advertising, where the focus is on revenue, not on facts. Happily it has become increasingly difficult for authorities to manipulate society through misinformation and censorship.

THE GREATEST CHANGE: PROGRESS BY THOUSANDS

Finally, on the list of most significant ways mobility and communication innovations have changed our lives is the description of a major change in the way technological progress is made. We now move forward by thousands instead of by one outstanding scientist at a time, progressing in ways reminiscent of the thousands who built the pyramids; such are the many people who made vital, though incremental, contributions to the accomplishments of the 20th Century. This change in the form of human progress is, however, anything but a camp of slaves driven by kings. Contributors were found within large and small corporations, uncovering innovation on their routine paths to new products. Others were ensconced in universities and research institutions, examining the fundamentals of science, when their moment of discovery arrived. In other places teams, even disparate teams, discovered the same idea or invention within days of each other.

Previous centuries' advances in science have most frequently been characterized by a cadre of names of those who forged new ideas in their minds, ideas that later changed the shape and face of scientific understanding. That list of great scientists has been published many times, one of the most recent in an excellent collection by Charles Murray, titled *Human Accomplishment*.

In contrast to the individual names, traditionally of such importance, the 20th Century's oeuvre, its output's most startling aspect, is that my four summary items under mobility and communication are the result of work and thought by literally thousands of scientists, engineers and implementers. And underneath the complete list and source for this book, are just as many thousands of others working and delivering devices, papers, prototypes, calculations, art, music and visions that eventually become a shortened list of three hundred and fifty major items that changed the way 20th Century society functioned, thought and moved. Even the way war was waged has changed. This increase to many thousands of contributors emboldens me to say that even the young man pumping gas in the '50s had a hand in delivering mobility to modern man. Human existence is becoming a human influenced mortar-work of a world; no longer a world-formed society but a society-formed world. We have been given more control over our environment, and so there is a need for us to exercise this control in an informed and responsible manner.

As humans have done through each major stage in social development, we will again absorb these new tools (as the orangutan did by using a bone to crush a foe in Kubrick's *2001: A Space Odyssey*) into the mainstream of life, invent a revised social order, and soon take this sped-up mobility for granted.

For now, the world is irrevocably altered. Every shirt we wear reminds us of Chinese labor's cost beneficence and they, in reverse, are reminded of American consumers who affect their employment. Every service call for assistance with a product's technical malfunction leads us to countries where English is a second language and service the first. In a recent inquiry regarding my cell phone bill I was routed to India for support, while a refrigerator inquiry was routed to California via the same "800" number prefix. Calls to a Chinese-American family's origins in the farthest reaches of China are made more easily than the delivery of high quality electricity to that same town. "Our town" is becoming more and more the Earth. In John Donne's eternal words, "If a clod be washed away by the sea, Europe is the less."

Reference notes

1. Photographic processes were invented in the 19th Century, but the specific instantiation used here was largely invented in the 20th.

CHAPTER 3

Table 3   The Physical Universe

CHANGE EVENTS	YEAR	CHANGE AGENT
Quantum Theory	1900	Max Planck
Invention of Light Quanta	1905	Albert Einstein — Ref. Planck
Special Relativity E=mc²	1905	Albert Einstein
Superconductivity	1911	H.K. Onnes
Atomic Nucleus	1911	Ernest Rutherford
Subatomic Behavior	1913	Niels Bohr
Star Lives — The Main Sequence	1914	Henry Russell
Proton	1914	Ernest Rutherford
Wave/Particle Duality	1923	Louis de Broglie
Universe bigger than Milky Way	1924	Edwin Hubbel
Matrix Mechanics	1925	Werner Heizenberg
Uncertainty Physics	1927	Born/Schroedinger/Heisenberg
Cyclotron	1930	Ernesto Lawrence
Neutron	1932	James Chadwick
e Spin	1933	Paul Dirac
Non-Local Realism	1933	Puldansky, Rosen, Einstein
Fission	1939	Hahn/Meitner; Strassman/Bohr
Cyclotron for Physics studies	1945	E.M. McMillan/V. Veksler
Hologram — Laser 3D Images	1947	Dennis Gabon
Information Theory	1949	C.E. Shannon
Anti-proton	1955	Emilio G. Segre/ Owen Chamberlain
Superconductivity	1957	J. Bardeen/L.Cooper/ J.Schrieffer/UI
Radio Astronomy	1957	Bernard Lovell
First Satellite — Sputnik	1957	USSR
First Rocket reaches the Moon	1961	USSR
Cosmic Background radiation	1964	G. Gamow (1916 Dicke)
High Temperature Superconductors	1964	K.A. Müller/J.D. Bednorz
First Rocket Landing on Mars	1971	USSR
Quarks — Elemental Particle	1968	H.Freidman/H. Kendall/ R. Taylor
First Person on the Moon	1969	USA Kennedy

"We have revealed the universe at the subatomic and star/galactic levels."

PHYSICAL UNIVERSE AND EARTH HISTORY

KNOWLEDGE OF OUR PHYSICAL UNIVERSE

The items listed under the categories of Physical Universe and Earth History can be summed up by four changes in our self-perception and our awareness of the larger universe and the microscopic world beneath us. The effect of this revised self-knowledge penetrated our psyches, affecting the personal reality of the planet's population.

1. Secrets of Matter: The century revealed detailed knowledge of the innermost secrets of matter. We understand the basic building blocks of all matter. Only gravity, of the many measurable forces, remains to be understood.

2. Global Connectivity: We have become a truly global society with satellites that provide weather, telephone, television and global positioning information for anywhere on the planet. There is no place on earth that cannot choose to overcome its isolation.

3. Earth in Perspective: Human beings now see their planet in perspective with the wider universe. We are aware of the planet's limited resources and its limited ability to cleanse its air and water. We have a dawning awareness of how far back human presence on Earth began, how races evolved in different places and what early man and woman were like. For the first time, the 20th Century has provided us a firming up of our sense of place on the long line of time. Just as early oceanic explorers guided their ships onto the Atlantic Ocean, so we begin our journey into the seas of our solar system and its universe.

The tremendous Saturn V rocket used in the US Apollo Missions 1967.

4. **Global Destructive Power:** All human beings are now aware of their nuclear weapons' destructive capability to massively and severely alter life on earth.

Attempts to understand the physical universe have occupied and bewildered human minds since records were first kept in Egypt four and a half thousand years ago. What has happened during these last hundred years? We have grappled with extremes of size and scale. We have revealed the universe at the subatomic and star/galactic levels.

To the person on the street, most of these major discoveries may not seem to have either personal or far-reaching relevance. My belief is that they do have that relevance, and, in fact, that even more obscure discoveries beyond those listed provide a necessary technical foundation upon which practical day-to-day convenience devices are dependent. In addition to convenience, these discoveries establish as different a worldview for each of us as did Galileo's Sun-centered solar system.

An illustration of the connection between obscure technical discoveries and familiar products in wide use can be found in the linkage between quantum mechanics and the silicon transistor (which forms the basis for all electronic devices in use today). A deep understanding of both the chemical and crystalline subatomic behavior of materials was required before the transistor could be invented. That understanding had as its basis the quantum theory and its offspring, all of which began with the 20th Century. All of the later subatomic discoveries — proton, anti-proton, neutron, electrons and their spin, the uncertainty principle, wave/ particle duality — had to occur before a more complete understanding of

Hubble Space
Telescope Image
of a Remote Galaxy
provided by NASA

material properties (related to surface cleanliness at the atomic level) allowed the first transistor to be made and tested by Shockley and company. And later, as mentioned in the preceding chapter, the absolutely critical invention of a means to create mono-crystalline silicon through the use of zone refining was required to create and exploit the highly useful silicon transistor and its later family members — computer processors ('Intel Inside'), global positioning receivers, cell phone transceivers and the whole panoply of audio electronics.

Secrets of Matter

MRI imaging is linked to our understanding of low temperature quantum physics through the fact that electrons don't bounce around as much when things are cold, thus reducing the resistance of materials to electron flow through them, turning them into superconductors. Late in the century the use of these super-conducting (low resistance) magnets for MRI (magnetic resonance imaging) has enabled viewing of the living body more thoroughly than ever before. The MRI machine provided a vastly improved, non-invasive way of obtaining diagnostic information about a person's internal condition. Without the very high magnetic fields of MRI magnets we couldn't get the electrons in our tissues to jump from one orbit to another, thereby sending out radio frequency signatures for different tissues that can be read on radios. Only dissection, a somewhat destructive way to investigate the human body, provides more detail.

MRI is a technology that, as we will see in the health chapter, is an essential and powerful tool for body investigation and repair. The cold-room temperature superconductors invented late in the century have not yet begun to change our electrical and magnetic tools, but they hold the hope of reducing the wire cross-section needed to create magnets, potentially making MRI as well as magnetically levitated vehicles easier to build. Super conductors at room temperature might save up to 80% of the losses caused by distributing electricity on overhead high-voltage lines, multiplying available electricity by half again without building one added power plant. This is not a discovery on our threshold, however.

Global Connectivity

My third illustration of fundamental physics affecting our lives can be found in our use of the ideas and mathematics behind general relativity, matrix mechanics, rocket engine technology and Shannon's information theory. All of which provided the knowledge base needed to successfully complete missions in space, and subsequently to launch communications satellites that enable data/voice reception and transmission to and from anywhere on the surface of our planet. Shannon's mathematical work on the theoretical limits of information content within certain radio bands became essential to defining the communications bands regulated by government agencies. Several of the newest cell phone technologies emerged based on his understanding of ways to increase the number of voice signals contained in one cell phone band. These cell phone technologies go by mysterious names like *spread spectrum* and CDMA (code division multiple access). The hardware that implements these theoretical methods of encoding and decoding our voice in order to transmit it to other places determines the cost of a call — so many cents per minute of a certain bandwidth. Other non-communications satellites provide the means for measuring the atmospheric effects of industry on our planet. It will be those measurements that settle the now heated discussions regarding the rate and causes of global warming.

Earth in Perspective

Humanity's vision of our place in the world has been vastly altered by the events and inventions in the physical universe list: surveys of our moon, Saturn's moons and Mars, seeing Earth from space, the discovery of star sequences describing a star's life cycle, measurement of the actual size of the universe by Hubbell, and finally the birth of radio astronomy and its use in the measurement of cosmic background radiation to look for the origins of the Universe. Our new sense of humility helps us accept our responsibilities as stewards for our Earth see it as a space ship with limited resources. That focuses us on research and development projects that can control pollution and pollution's negative side effects on our living conditions.

Physics has also given us tools needed to measure the atmospheric chemistry. The same understanding that provides the knowledge of the atom and light quanta also provides the absorption spectra used in assessing the constituents of the gasses of Earth's atmosphere.

Global Destructive Power

And of course, as in all human endeavors, these physics-based discoveries provided knowledge that led to its use in creating new weapons. Understanding the atom, and Einstein's energy-relationship to atomic mass, opened the door to the creation of the atomic bomb, a weapon capable, according to some scientists, of permanently damaging human life on Earth. This has, for the first time, forced humankind to set limits on man's destructive decisions. We must not go too far. A rocket-delivered atomic bomb can place very large explosives, with very long radioactive after-lives of destruction, anywhere on the planet.

20th Century advances require us to manage the awe-inducing power of the atom and the transistor as wise caretakers of our people and planet. The coming century will surely ask for even more wisdom in managing our coming scientific achievements, such as the need to control cloning of people themselves.

Science and social science will not let us rest nor allow us to avert our eyes from the responsibilities placed on us for using good judgment. Nor can we believe, as some propound, that regression to simpler minded times is a solution. There is no single point to which we can pull back from this brink of knowledge in order to save ourselves from the need for responsible behavior. We must plunge into the pool of knowledge gained and find our way as Adam and Eve did after tasting the fruit of the tree of knowledge, using the best tools of every epoch.

· · · · ·

Table 6   Earth History

CHANGE EVENTS	YEAR	CHANGE AGENT
Radioactive Decay (Carbon) Dating	1907	Bertram Borden Boltwood
Origins of Continents and Oceans	1915	Alfred Lothar Wegener
Automotive Smog — Los Angeles	1946	R. R. Tucker/Haagen-Smit
Int'l. Geophysical Yr-Atmos. Monitoring	1957	World governments
Ocean Floor Topography-Plate Tectonics	1960	W. M. Ewing and Columbia U.
Chimpanzee — tool use similar to human	1960	Jane Goodall
Australopithecus — Homo Habilis	1961	Louis Seymour Bazett Leakey
Silent Spring	1962	Rachel Carson
Acid Rain	1967	Coal Electric Power Generation
Green Revolution-higher grain yields	1968	Norman Ernest Borlaug
Cleaner Exhaust Automobiles	1970	USA EPA Clean Air Act
Ozone layer impact of Freon	1974	Molina,Crutzen,Rowland-UCI
Documented Ozone hole in Antarctic	1985	Farman/Gardiner/Shanklin/UK
Montreal Protocol — Limit CFC's	1987	United Nations and scientists
Kyoto Protocol — Limit CO_2-Greenhouse	1996	UNFCCC

EARTH HISTORY

As mentioned earlier, this small list of the events of Earth's history also changed our awareness of our place in the Universe and on Earth. The discovery of carbon radioactive dating for living carbon containing materials allowed us to determine the age of various human skeletons. For the longer time-frames, zircon dating, among other tools, made it possible to find the beginning of Earth time and to be surprised by the number of thousands of millions of years old our Earth is (4.6 billion years old is the most recent estimate). This aged-earth awareness has augmented our understanding of Darwin's evolutionary theories from the preceding century, making the time scale for such astounding evolutionary changes more palpable. Followed then by the discoveries of continental drift and plate tectonics, we have become more knowledgeable about the land

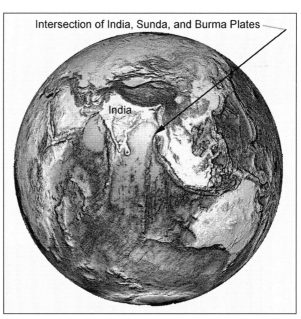

Topographic map of Earth's crust — plate intersection that caused the 2004 Tsunami resulting in a horrifying loss of life

origins of all peoples and their later migrations. As recently as 2004 those plates caused the largest earthquake and loss of life ever recorded. That earthquake in Malaysia killed two hundred and thirty thousand people following a plate shift that caused a large tsunami.

Toward the end of the century, we became aware that the industrialized nations' consumption of fossil fuels and fluorocarbons had begun to affect the entire planet. Rachael Carson in her book, *Silent Spring,* predicted a similar phenomenon involving pesticides two decades before scientists discovered the ozone hole in the upper atmosphere. Subsequent to her book, other planetary nightmares emerged to illustrate how human consumption was changing our planet. Previously, the earth had seemed able to accommodate our numbers and absorb our waste.

The 20th Century hypothesis that a CO_2 greenhouse effect might cause a discernible increase in average surface temperature, though still controversial, is another fear on the brink of becoming real, potentially affecting the future of millions of people living in now-habitable low-lands.

In the same 20th Century time frame, acid rain, caused by burning coal for electricity, began to harm our forests. Other industrial gasses emitted by automobiles — carbon monoxide and nitric oxide — forced eventual government intervention to reduce photochemical smog's primary visual and secondary biological effects. Our government mandated automobile design-control regulations have significantly reduced the impact of automobile pollution in North America, especially in Los Angeles, a city of six million automobiles and ten million people. These events also provide a warning to other nations, some who have yet to act. Their delay arises from slower industrialization and lack of government intervention, not from failed awareness.

We humans also collide with the side effects of regulation, of the technical changes made to gasoline to improve the completeness of combustion. Using the additive MTBE, we cleaned up the exhaust gasses only to find that chemical seepage of MTBE from gas station tanks into our ground water increased the carcinogenic risk to humans. MTBE has been removed from the fuels. We now know that making changes too quickly, without careful assessment using a pilot run (a carefully thought out small-scale experimental protocol that assesses secondary effects) may also create additional problems that the changes were intended to cure.

Summary

What do we ultimately make of these changes in our knowledge of the Physical Universe and Earth History? What is the kernel of relevance within this newly acquired knowledge? I suggest by restatement, the following:

Earth in perspective

We see our place alongside other life forms, the many plants and animals, all cohabitants of this chemical Earth, which we now perceive as an infinitely small speck on the vast landscape of space and time. This awareness of scale can sometimes frighten us with its seeming implication of our insignificance. Yet it also tells us that earth is a very special and unique place, which, even if we are not both honored and obligated to protect, we must do so to survive. Our neighbors in the undeveloped world are so close, so much a part of us, we must find ways to work together in better balance, a federalism if you will, where different forms of government and life styles are allowed to co-exist in a tense harmony.

Instant global connectivity from the secrets of matter

This connectivity was discussed extensively in Chapter 2, though the science that enabled it was space travel and our ability to place satellites anywhere above the earth. And our knowledge of the inorganic world allows us unprecedented manipulation of the elements that submit themselves, sand, copper, etc., to our machines — computers, music, MRI machines and more.

Global reach of knowledge and destructive power

Our understanding of the structure of all matter provides us access to almost infinitely durable electric power as well as massively destructive weapons, whose destructive power makes them almost unusable.

Fear of catastrophe — its exploitation by governments and media

Our late 20th Century was marred by exploitation of our fear regarding our potentially catastrophic human impact on the earth. Catastrophe was prominent in movies, in books, in the news media, and in the literature of the environmental protection movement. Our own information-conveying organizations have also preyed on us without discretion or evidence. Never was the actual data behind the Kyoto protocol and the reasons the U.S. did not sign the treaty revealed, only the implication that the U.S. was ignoring the consequences of global warming. Did those misrepresenting the information describe the economic consequences of compliance with the protocol to consumers? Did the presenters describe what the per-capita CO^2 reductions were based on and how differently large industrial nations would be impacted compared with countries like China with a population of two billion and only two million automobiles? Do the UN science teams have proof beyond dispute that industrially generated CO^2 is the cause of the warming?[1]

We must develop the ability to present the science behind each worst-case scenario more completely, and also to present the best-case scenario if we are to survive emotionally and rationally. Constant reversals of reported science based outcomes frustrate those we wish to keep informed and devalue the credibility of our media.

New Tools

The new tools of Carbon 14 and Zircon radioactive dating provide ways to measure information about the Earth and humans across wide spans of time, from thousands to billions of years. The ability to establish an accurate date in the past for the revelations of geological, biological and archeological detective work has provided us with the first accurate time-line for earth and the stages of biological evolution. With these tools we can use forensic work to establish the sequence of changes along the paths of plant, animal and mineral histories. We are only now establishing baselines that help in assessing human impact on our environment, without which no plan can be made for our future.

We have gained a growing sensitivity to our planet and our need to manage our use of it. The first corrections have been made using technology to control the impact of human use of the planet and there is an increasing awareness of the complexity involved in achieving that control. We have a lot left to do. [2]

That is enough.

Reference notes

1. IPCC Working Group One, a UN scientific body, published the 2001 report at: www.grida.no/climate/ipcc_tar/wg1/index.htm.

2. Choosing coal for electricity has added CO_2 gasses to the atmosphere, potentially causing some of the increase in global surface temperature. We closed down nuclear power generation because we were afraid of the ghost of undefined consequences;

Class of 1998 astronauts below Saturn V rocket's engines

Reference notes (continued)

radiation. Nuclear power, however, does not generate CO_2. That is a good thing. We must be less catastrophe oriented in our decision making and more scientifically and data driven.

CHAPTER 4

Table 4   Life Utilities

CHANGE EVENTS	YEAR	CHANGE AGENT
Safety Razor	1901	King Camp Gillette
Air Conditioning, temperature control	1902	Willis Haviland Carrier
Invention of the Electric Utility	1903	Thomas Alva Edison
24-gear Mountain Bikes	1910-1987	P. de Vivie/T. Campagnolo/ Tour de France
Bakelite (first fully synthetic plastic)	1910	Leo Baekeland
Home Refrigerator (Domelre)	1913	F. Wolf/A. Mellowes
Frozen Fish Process	1923	Clarence Birdseye
Electro-static Precipitator — particulates	1923	Fred Cottrell
PVC (Polyvinyl Chloride) Modern Plastic	1927	F. Klatte(1912), Waldo Semon
Scotch Tape	1929	Richard Drew/3M
CFC = Chlorofluorohydrocarbons-non-corrosive fluid	1930	Thomas Midgley Jr.
Gold Standard for currency abolished	1933	U.S. Government
Computers/Programs/ Artificial Intelligence	1937	Alan M. Turing
Nylon-synthetic fabrics	1937	Wallace H. Carothers
Ball Point Pen	1938	Laszlo Biro

Table 4   Life Utilities (continued)

CHANGE EVENTS	YEAR	CHANGE AGENT
Teflon-PTFE Neutral High-temp plastic	1938	Roy Plunkett
DDT= Insect/Typhus disease control pesticide	1939	Muller/Ziedler
Fission Reaction/Nuclear Electric Power	1939	Enrico Fermi
Duck(duct) Tape	1942	Johnson and Johnson Corp.
Scuba Aqua Lung	1943	Jacques Cousteau
Microwave Oven	1947	Percey L. Spencer
Velcro (idea from a plant)	1948	George de Mestral
Sedatives	1952	Robert Wilkins
Composite Carbon-fiber, strength-weight	1950s	Rolls Royce/Royal Air Force
U.S. Interstate Highway System	1953	US President Eisenhower
Electronic Quartz Watch	1968	Centre Electronique Horloge
Liquid Crystal Display LCD	1971	RCA, Kent State, LaRoche
Pocket Calculator	1971	Texas Instruments Inc.
Post-it Notes	1980	3M corp.-S. Silver/A.Fry
CAD (Computer aided design) Software	1982	Many contibutors
Nuclear Reactor disaster — Chernobyl	1986	USSR

"This new engineering maturity has made it possible for small numbers of people to produce millions of units of any device."

LIFE UTILITIES

A surprising number of the significant and influential discoveries, products, inventions and events from the 20th Century had their impact on practical day-to-day convenience — our Life Utilities so to speak. There are those who argue that these innovations did not significantly alter human development or life on the planet, that they were just another adjustment to the cook-stove, the washing machine or the dinner table. Some certainly did not change life much, such as the microwave oven. On looking more closely, however, might we reassess the generality of that quick, skeptical judgment? Let us see.

Before delving into the Life Utility list in detail I will share my own conclusions about the thousand year impact of two new characteristics of post-industrial society:

1. The Maturing of Design and Manufacturing Technology

The maturing of design and manufacturing processes that provide highly complex products, optimized to meet consumer and society needs, is a recent and astounding phenomenon. This maturity allows costs, durability, function and beauty to be managed during the design phase, before the high costs of tooling and setting up production are incurred. Mass production of a new design provides people access to products largely independent of affluence because the component costs are so low. Distribution costs tend to dominate price.

The magnitude of this change is similar to the advantage gained with the combine for harvesting grains, enabling the feeding of large populations; or of the cotton gin in an earlier century, which automated and streamlined the conversion of cotton to cloth. This new engineering maturity has made it possible for small numbers of people to produce millions of units of any device.

2. The Imperative of Energy

The creation of the electric utility company with its generation and distribution of electricity, combined with the broad-based availability of household appliances (which includes air conditioning and forced air heating) and the mass use of the automobile, all recent Life Utilities, formed a new life-essential hunger in modern society — the desire for energy products — gasoline and electricity. In the 20th Century, sources of energy to run machines became as vital to a family as food.

To substantiate these conclusions, let's review the fairly short list of Life Utility listed items to see how these sweeping changes to society began.

MATURING OF DESIGN AND MANUFACTURING TECHNOLOGY

This first group from the Life Utilities list illustrates the critical elements of a maturing design and manufacturing technology that, by the end of the century, had revolutionized the process and reality of new product creation.

Model -T Automobile Assembly Line 1908 Henry Ford

With the advent of mass production and interchangeable parts, the automobile industry of the early 1900s came face to face with tradeoffs in design between durability and manufacturing cost; i.e. profit. A tradeoff to an engineer occurs when two desirable attributes of a design cannot both be used in the same design because something about them is mutually exclusive. This is not always obvious. Just because it hasn't been done before does not always mean it can't be done. A tradeoff must be made, that is a trade between one desirable aspect of a design called a requirement and another requirement, when there is no known technology that allows both things to be designed into the same gadget. For example; a flashlight that is small, light weight and whose batteries will operate for a long time (say a month of all night operation) are mutually exclusive at the present time. A tradeoff between weight and operating time must be made and is influenced by the convenient availability of D cell batteries.

As the science of automobile materials and wear mechanisms matured, the durability vs. cost alternatives became more a matter of design choice, time-to-market, design complexity, material used, and manufacturing cost. Previously these tradeoffs had been less a choice from a basket of possible materials and methods and more a result of the limitation of each craftsman's training, experience and vision. In past centuries when design and manufacturing were crafts learned from master teachers, durability was achieved as time and experience made improvements obvious.

Two examples of 20th Century design-tradeoff choices rather than craft-experience choices illustrate this difference.

First: The Mag-Lite, an extremely durable flashlight costing 3 to 4 times that of the standard battery manufacturer's flashlight which lasts no more than a year. The Mag-Lite founder made the following design and manufacturing decisions that determined, *a priori*, its durability. The Mag-Lite features an all metal housing, anodized for appearance and ruggedness with O-ring sealed ends and screw-on caps for both the light chamber and battery in order to create a rugged, moisture resistant design. These flashlights now populate the landscape, eliminating yearly purchases of cheaper replacements.

Second: The operating life of a mainframe banking computer cooling fan, formerly limited by its shaft bearings, was increased by choosing ball bearings costing ten times the price of those used in off-the-shelf fans. The improved bearings eliminated hardware failures in that subsystem until that computer became obsolete seven years later. The choices were made to meet a specific up-time or reliability target for the banking industry and were economically successful in that application. The fan price more than doubled to two hundred dollars.

Safety Razor   **1901 King Camp Gillette**

Disposable shaving blades might have been industry's first move into the manufacturing of short-term disposables, which later in the century became one of the most lucrative product types because of their predictable replacement demand. As these and many other replaceable components were discarded in ever increasing numbers, the United States became known as the "throwaway society." The economic engines of industry benefited from the constant demand for new razor blades, sandwich bags, plastic cups, hospital needles, etc. These disposable products, which started out looking great for ease of use and steady demand, began to stack up in garbage dumps. As the public became more and more aware of our limited supplies of oil, clean water, and living space, the concept of recycling became more vital to our planned survival. Thus emerged our government-driven recycling industry. More about the tradeoffs related to disposability later.

Bakelite (1st synthetic plastic)	1910 Leo Baekeland
PVC (Polyvinyl Chloride)	1927 F. Klatte(1912), W. Semon
Nylon	1937 Wallace H. Carothers
Ball Point Pen	1938 Laszlo Biro
Teflon (a high-temp plastic)	1944 Nuclear Bomb Studies
Velcro	1948 George de Mestral
Composite Carbon-fiber	1950 Rolls Royce/Royal Air Force
Post-it notes	1980 3M corp.-S. Silver/A. Fry

Synthetic plastics (Bakelite, PVC, Nylon, Teflon) and Graphite composites, and their use in molded forms gave birth to low-cost, mass-produced assemblies, especially for electronics. They also became extensively used for clothing, containers, eye glasses, contact lenses, kitchen utensils, bowls, aesthetic covers for all motorized appliances, light fixtures, toothbrushes, measuring cups, pipes, fasteners and construction laminates. Plastics brought easy access to utility-based tools for people in most social classes worldwide. Plastics are essential to the cost reduction of complex assemblies that use electronics. Our CD players have fiberglass/plastic circuit boards, plastic encapsulated silicon ICs (integrated circuits), plastic laminated motors, switches, transparent windows and keypads, housings and sealing rings. The ability of chemists to formulate their flexibility property, impact strength, melting temperature, flow characteristics and colors, made them critical components in the maturing of design and manufacturing technologies.

Velcro is an easy attachment method for shoes, jackets, pockets, and flexible material containers. This material has become the single most used means of cloth attachment in all its varieties from tennis shoes and back packs for the young to enabling elders who have diminished dexterity to put on shoes and coats without help. More importantly, velcro is one of many of our more evolved engineering designs, a more organic and more integrated solution than tie-down devices requiring good hand-eye coordination. Velcro provides more than double the life of a shoestring or button system due to the inherent reduced wear mechanisms of the small plant-like hooks when compared to shoe eyes and hooks that abrade shoestrings. Similar to crash resistant bumpers and vinyl siding for homes, these more evolved engineering designs using advanced materials provide longer operating life and reduce the amount of material discarded during a lifetime. This is not a massive phenomenon in industry at present, but a factor people can use to make decisions regarding cost and operating life — to direct their vote (purchase), as it were, toward our future.

Attribution Discussion

Before I start the discussion of the next items on the list, I'd like to elaborate on the difficulty of assigning names to the items on the lists. The LCD (liquid crystal display) illustrates the complexity of attribution taking the broadest view of an invention. LCDs are the dull grey screen on your calculator, cell phone, and even the back lit screen on your computer or your flat screen TV. While trying to put *one* name in the box for LCD, I was corrected by a scientist close to its inventors and their company. Here is the revised sequence of invention as I understand it now:

The idea of using a liquid crystal as a display technology was stimulated by a paper (published in the journal *Chemical Reviews)* by Chemistry Professor Glen Brown, PhD, 1956, and one of his graduate students, W.G. Shaw, both at the University of Cincinnati. The paper summarized the then known science and chemistry of liquid crystals, making the subject readable to other chemists, scientists and engineers. Only nine years later, in 1965, at RCA, David Sarnoff's scientists, Richard Williams and George Heilmeier, published a paper on the possible use of liquid crystals as a display technology. They then proceeded to fabricate devices and by 1967 had demonstrated liquid crystal's display performance. Those LCDs were power hungry and therefore not yet the breakthrough needed for use in mass production electronics. In the late 1960s, two separate groups of researchers found an alternative formulation of the crystal that required much less power to operate. They were James Fergason at Kent State University in Ohio and Martin Schadt with Wolfgang Helfrich, working in Switzerland for F. Hoffmann-La Roche. These three scientists in two isolated laboratories created the first of these low power LC displays in 1969 and 1971 respectively. Within five years the vision formulated at RCA had become a working, desirable reality. The first calculators using LCDs in large quantities were available by the late 1970s; 1976 through 1979. LEDs had previously dominated the calculator readout market.

Now to answer the question I began with; which name or names should be shown as inventor? Take a second look at the answer on the next page. In the end, are the names themselves most critical or are they really symbols given for naming the technology itself, the LCD? Because of these attribution vagaries I will at times choose an inventor and at other times an implementer. Of course the names are critical to the circle of inventors around a new idea, but often it is a team, or more an army of inventors, most of whom never get the public visibility they deserved.

Onward:

Computer Programs/	1937	Alan M. Turing
Artificial Intelligence		
Liquid Crystal Display (LCD)	1970	RCA, Kent State, LaRoche
Pocket Calculator	1971	Texas Instruments Inc.
CAD (Computer	1982	Many contibutors
aided design) Software		

Alan Turing is remembered as the inventor of the idea of a computer and its first implementation. He envisioned a calculating machine that had a stored "program" of intermingled data and instructions. His calculator would use the instructions to crunch the data into new forms needed by people in science and business — one computing machine, one memory, essentially the computer we use today. From those roots and combined with thousands of inventions and implementations, modern computers later merged with the software of computer aided design tools (CAD) in the 1980s. Many inventors contributed parts of the software system needed to render two dimensional and later three dimensional drawings. [1] Still later, CAD tools began to provide unprecedented analysis of engineering designs, allowing the

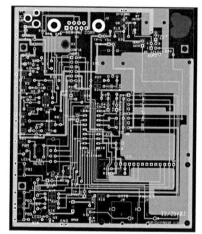

Printed circuit board

visualization of fluid flow for molded plastics. This made it possible to: optimize tool design, reduce the stress patterns within plastic or metal parts, predict and minimize failure points in bridges, cell phones, even rockets. These tools also provided automation of the routing of photographically etched electronic circuitry on the multiple layer "printed" circuit boards (some having hundreds of thousands of connections) found in all electronics today. Eventually these routing tools provided engineers with the automation required to lay out microcircuit designs for computer "chips" using millions and millions of transistors. Without such tools these tasks would be beyond human reach. John Walker said at the founding of his company, AutoCAD, in 1982: "Computer aided design is the modeling of physical systems on computers, allowing both interactive and automatic analysis of design variants, and the expression of designs in a form suitable for manufacturing." The availability of high-speed, low-cost computers and the design demands of electronics extended the CAD frontier by placing pressure on all aspects of the tools and modeling techniques. Not until the 1990s did overall cost of computers and tools reach a level that made CAD design support ubiquitous across the design and manufacturing industry. These tools became capable of performing computational fluid dynamics as if they were elementary math, giving amazing power to their users and so forming another essential ingredient of the maturing of design and manufacturing processes used by industry.

THE IMPERATIVE OF ENERGY

The second group from the Life Utilities list illustrates key innovations in a century-long emergence of the electric power industry. That industry, combined with the mobility provided by our freeway and automobile society, solidified our need for energy in the planet's social-order. Let's review the list:

Invention of the Electric Utility	1903	Thomas Alva Edison
Electro-static Precipitator	1923	Fred Cottrell
Revolutionize Road Transport	1929	Autobahn — Germany
Fission Reaction/Nuclear Power	1939	Enrico Fermi
U.S. Interstate Highway System	1953	President Eisenhower

Neither the piston combustion engine nor the electric motor was invented in the 20th Century, but their availability to billions of people occurred only in the 20th Century.

The 20th Century availability of electricity was rooted in three life utility implementations: first the electric motor with its ability to provide motion at a remote location or to generate electricity from mechanical motion if rotated by some external force such as a diesel motor, steam turbine or a water fall; second, a means for distribution of electricity to far-flung homes and businesses — essentially the electric power utility — and third, appliances for heating, cooling, lighting and communication supported by the availability of electricity in our homes and businesses.

The distribution of electric energy, starting with Edison's *electric utility*, made electric power ubiquitous across the planet and became a signature of our unique century. The home went from three thousand years of wood and coal fired heating, and lighting by oil lamp, to predominantly electricity driven systems by the end of the century. The figure on the facing page illustrates the changing types of fuels consumed over the span of three and a half centuries.

Even the modern oil furnace requires an electric fan and pump to produce its heat. As demand for electricity rose, the increased burning of coal and oil led to pollution near power plants. The invention of the electrostatic precipitator (and other cleaning systems such as scrubbers to remove Sulfur compounds) provided a means of removing the polluting particulates and other undesirable contaminants from power plant smoke stacks.

The decisive final links to an expanding energy-use profile were provided first by the mass-produced model-T automobile, eventually coupled with Eisenhower's highway system in post World War II America and the Auto Bahn by one of its many names in Europe. Later, commercial airlines increased fuel consumption with thousands of flights around the globe. The energy market became the fourth largest economic force[2] on the planet, narrowly behind such stalwarts as medical care, housing and food.

Coal, oil, natural gas and nuclear fuel all weighed in over the span of this past century as the best source of both electrical, heat and vehicle power. In no previous century was so much energy needed or used. There have been, of course, downsides to the energy imperative: in 16th Century London coal soot blackened her skies. Each fuel has its "soot:" coal its CO_2 and acidified rain; oil's sulfur and nitric oxide smog; nuclear fuels damaging and long-lived radioactive waste products (in spite of yielding the greatest energy per pound of fuel, meaning its waste products are far smaller than those of coal or oil), and finally, innocent natural gas whose only failings seem to be its piping costs and resource limitations. Hydropower, using a dam's waterfall energy, makes up a small part of

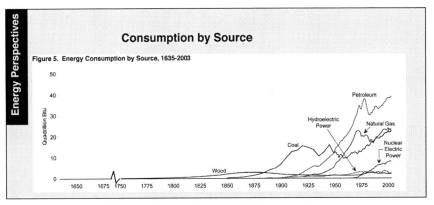

Historical growth of US Energy Consumption from 1650 to present comparing sources (early time scale compressed at jog)

energy in many countries and is renewable. These dams tend to consume natural canyon landscapes and have been unpopular lately. The solar and wind power contributions are so small (see bar chart on next page) as to not deserve mention but for their support of low power requirements at remote locations.

Electric power generation using atomic energy emerged as a hedge against the exhaustion of fossil fuels (coal, gas and oil) and was, to a large extent, invented by Enrico Fermi, an Italian physicist in Chicago, who introduced this peaceful use of a sustained nuclear reaction to the world in 1939. He imagined and invented a working power generation scheme using a thermopile as generator and a nuclear reaction as a heat source. He was the first to control the same nuclear reaction used in the atomic bomb so that it generated peaceful electricity. Later, in 1951, sixteen scientists and engineers — all staff members of Argonne National Laboratory — designed and built the first electric power generating reactor, EBR1, and recorded their historic achievement by chalking their names on the wall beside their generator. The reactor could be shut down as needed, and controlled in a linear manner by reducing the reflection of neutrons back into the reactor, thus controlling the quantity of new neutrons spun off the core of the uranium 239. Controlling the rate of the nuclear reaction provided control of the water temperature. These 1951 basics are unchanged since nuclear electricity's inception, though the details have changed. Modern nuclear reactors are safer for many reasons: the strength of the containment vessel, better control of fault conditions, the use of natural failsafe feedback to eliminate runawy reactors, the choice of materials and finally by reactor core design itself. Radioactive waste can be made less of a problem in some experimental designs (MOX breeders) where the half-life of the final plutonium is reduced to tens of years rather than thousands.

To give you a sense of how much nuclear power is present in modern electric power generation worldwide, I quote from a summary of the top consumers and their electric power generation source dependence

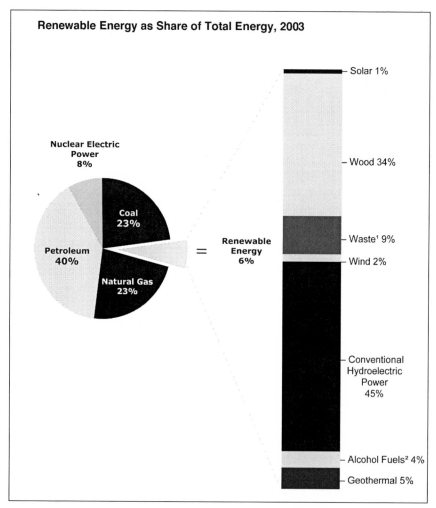

Renewable energy (percentages are of all energy sources) illustrating how small solar and wind power are compared to other more effective renewable sources of energy.

on nuclear energy: 20 percent of the electric power in the United States is generated by nuclear reactors, 80 percent in France and Holland, 50 percent in Sweden, the most safety regulated country in the world. Germany and the United Kingdom use nuclear for 25 percent of their generating capacity and Japan uses nuclear for 53 percent of its generating capacity.

The availability of nuclear electric power provides a safety net from the loss of fossil fuels. This loss may occur sometime in the next 300 years (Opinions vary so widely that some predict 100 years and some 1000!)

Dire mid-twentieth-century predictions were made of exponentially increasing electric power consumption in the United States. That doomsday scenario evaporated around 1975 when demand flattened. For example, late in the century the growth in demand for electricity had slowed so significantly that no new power plants were built in California between 1985 and 2000. The doomsayers were temporarily silenced. We now understand from careful perusal of data, that per capita use of energy tends to saturate or flatten as one reaches a comfortable plateau in the quality of life, such as that of industrial Europe and North America.

· · · · ·

As we stated at the beginning of this chapter, there are two significant aspects to the developments of the 20th Century with regard to Life Utilities:

1. The Maturing of Design and Manufacturing Technology

The maturing of design and manufacturing processes that provide highly complex products, optimized to meet the needs of consumers and societies, is a recent and astounding phenomena. This maturity allows life-cycle costs, durability, function and beauty to be managed during the design phase, providing people access to products largely independent of affluence. This maturity made it possible for a small number of people to produce millions of units of any device. The benefit to society in use of this tool is up to society itself specifying well formulated requirements for their products.

2. The Imperative of Energy

The creation of the electric utility, combined with the broad-based availability of household appliances (including air conditioning and forced air heating) and the mass use of the automobile, all recent Life Utilities, are at the root of our desire for energy sources such as gasoline and electricity. In the 20th Century, these sources of machine energy have become as vital to a family as food for their bodies.

We've reviewed the first level of detail on Life Utilities and we saw how they linked from the two society changes summarized to the individually listed items. I ask your indulgence in my continued expansion in this next section as I elaborate on these two society changing arenas, Manufacturing technology and the Imperative of energy. They are at the heart of the tension between conservation and human industry. Through these two stages of discussion you will see much more clearly where we were at the end of the 20th Century.

THE MATURING OF DESIGN
AND MANUFACTURING TECHNOLOGY

Over the span of the 20th Century, corporate understanding of a product's complete life cycle grew to encompass its existence and economic viability at all stages: from conception of an improved or new product — through the design, prototyping, testing, manufacturing, and revising where needed (based on field feedback) — to the product's maturity as a commonly purchased, high demand, profitable commodity product, and finally, to the product's death. During the design stage of a product, cost and profit were, until recently, the only objective. By the end of the 20th Century other parameters had become essential and were added to this cost-only formula: durability, aesthetics and society cost (e.g. non-renewable resource consumption, scrap cost, fair employee working condition costs). They are now all part of the total design requirement. Engineers and chemists are asked more and more to design toward a reduction of long-term negative environmental impact costs as well as users' needs. Regulatory agencies became the means of managing these social costs. Their regulatory purpose was to burden each manufacturer equally with society-critical design requirements intended to optimize a design's benefit to an individual and to manage the societal cost in terms of resource depletion, safety, and eventual disposal.

One of the most infamous examples of society's limited foresight in the management of long term social cost can be found in the assessment of atmospherically destructive automobile combustion by-products. Another example involved the early assessment of high speed automobile passenger safety. In both cases fixes were generated some 50 years after first introductions of the automobile. The catalytic converter came about as a result of regulator mandated reductions in NO (nitric oxide) combustion byproducts, while the seat belt and air bag were the result of legislated safety requirements. Regulators can both impede and enable development of new products, at times over-regulating with a subsequent loss of progress and increase in expense. I believe that in the long run (one hundred years), competition between companies ends up balancing the forces of commerce, product cost and regulatory requirements and tends to move products steadily toward greatest usefulness and minimum overall social cost, smoothing out regulatory impediments where they exist and adding them where they are absent.

As mentioned earlier, the advantage of a disposable product is its stable production demand. The disadvantage, of course, is increased non-recyclable garbage.[3] The disadvantage of a durable product is the eventual saturation of demand, which severely reduces production quantities and increases cost due to the *dis-economies of scale* (the increased cost of producing fewer of each item making the setup and tooling costs become a larger burden for each item). These qualities, disposable vs. durable, form two endpoints on the product life continuum and are two of the primary social/industrial tradeoffs constantly in play within a manufacturing society. The choice about how much durability to build into a product is driven by many factors — cost, cost to repair, the rate of change of technology and the desire to just change things or get new things.

And so it is, our improved design and manufacturing technologies enable us to use our engineering skills to design for a product's total life and to optimize among the many variables; user desired functionality, society's management of waste, overall cost to society, and responsible use of the planet's resources. Society must take responsibility for providing the specifications for making these tradeoffs. This responsibility must be filled, not by more government agencies, but by individuals through our buying habits, voting with our money and our feet, just as we are doing with the Mag-Lite and its disposable partner, the Life-Lite.

THE IMPERATIVE OF ENERGY

The necessity for energy in modern society is best illustrated by looking at who uses it and how much they use (electricity, gasoline,

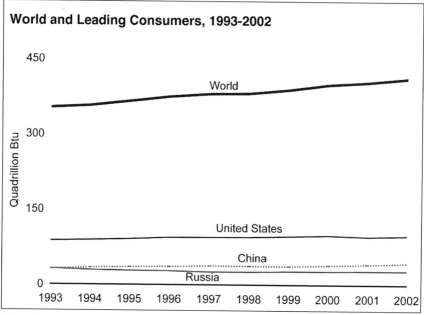

World and Leading Consumers, 1993-2002

Illustrates recent worldwide growth in energy consumption with some interesting detail.

natural gas, heating oil). In the United States, as of 2003, transportation (largely automobiles) consumed 27%, residential electrification 21%, factory electrification 33%, and commercial (office and store-front) used the remaining 18%. That means we as individuals used 48% of the total energy burned in the United Sates for direct personal purposes (27+21).

We have met the enemy and he is us.[4]

The United States is the best example of a worst case consumption scenario. Energy resources consumed in the USA[5], starting in the year 1800 and ending in 1900, went from half a quadrillion[6] BTU (or QBTU) to 6 QBTU, a growth of 12 times the energy use — the early 19th Century fuel being predominantly wood, with coal coming in second. Over the span of our 20th Century the predominant fuel changed several more times, first from wood to coal, then to petroleum, then to natural gas and finally all types became end-use dependent, the automobile being the prime consumer of gasoline. Home heating was split between natural gas and heating oil and commercial electric power generation was dominated by coal, though with significant nuclear support. By the end of the 20th Century (2002) U.S. energy consumption rose to 100 QBTU/yr, another gain of 12 times the energy use. Worldwide use in

Human light on Earth in the late 20th Century

2002 was 412 QBTU/year, a growth of 16 times the energy use at the beginning of the century. The worldwide population grew only 4 times its 1900 size in that same time span. Africa and the Middle East were the low consumers at 13 and 19 percent, respectively.

This late 20th Century image of the earth from space illustrates most vividly that energy is equal in importance to food. This image is the single most powerful statement of human civilization's dominion over our Earth and of our responsibility to take care of it.

This same photograph, if it had been possible in 1900, would have shown a largely dark world, the muted dimness of gas and oil lamps nearly invisible compared to the light visible in this 1990's NASA image from space, and these lights consume only one tenth of the electrical energy used each day and only one hundredth of the total daily energy use.

For the coming decades, possibly centuries, traditional fossil fuels will be sufficient to support the energy needs of Earth. After that, there will only be nuclear and hydroelectric. Despite optimism and popular culture's desire, corn and wood take too much land area to support the energy appetite of a highly populated Earth. Other renewable sources — wind and solar — are, and will continue to be, such small contributors that they should not be considered (except of course where remote location makes it financially preferable due to lack of power line infrastructure), given the size of the world stage. Sadly, the United States shut down all nuclear development sometime in the 1980s. This was the direct result of public fears, first involving the unsettled question of where to put the thousand-year rate of decay of radioactive waste products, and second, the fear — founded or unfounded — induced by the failure of two nuclear plants' safety systems. Only one of them, Chernobyl, caused significant immediate human loss of life (31) when it overheated and melted its radioactive core, eventually forcing the release of radioactive gas and airborne particulates. It is believed to have led to thyroid cancer in 700 children or more over the 6 years following the accident. Estimates by WHO suggest the number of cancers might reach 4 to 8 thousand, but true quantities using causal analysis will continue to be difficult to assess as pointed out in the 1996 WHO report.[7] (Added note for perspective: Most thyroid cancers are papillary thyroid cancer and this is one of the most curable cancers of all cancers that humans have.) If 100 people die every day from breathing coal smoke, it doesn't make the headlines. But if a single rare nuclear event takes 10 lives, it becomes a huge deal in the public consciousness. The greatest harm caused by Chernobyl was the

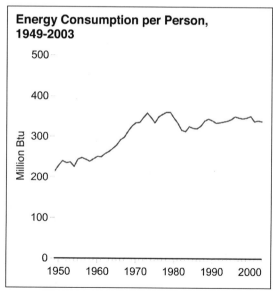

Energy Consumption per Person, 1949-2003

Flattening of per-capita demand for energy in the U.S.

displacement of several hundred thousand people from their homes to get them away from the radioactive ground.

On the positive side of this "is nuclear safe" question, there are over 439 plants now in operation worldwide, with a total of 38 million cumulative hours of successful power generating operation. If the public assessed highways and automobiles with the same fearful tentativeness, they would have stopped building cars many years ago. Electric power will continue to use primarily non-oil based fuels, and therein lies the comfort of staying with fossil fuels. On the other hand, it is imperative that we begin carefully implementing different means of handling nuclear waste to gain experience based on calculations and experiments, which have already indicated the safety of two recommended disposal means; passivation and deep burial. Fearful NIMBY (not in my back yard) behavior — avoiding these developmental reactors — will not lead us to perfected designs when the need for nuclear energy becomes critical.

An extremely interesting aspect to energy growth is that its per capita usage (lighting, heating, refrigeration and transportation) flattens out quite nicely once a certain living standard is reached (see figure on opposite page). Industrial and commercial uses also tend to flatten on a per capita basis. Since United States growth is at present primarily fueled by population growth driven mainly by immigration, there is continuous energy consumption growth indicated for the new century. Growth worldwide can be expected to follow the 1930-2000 pattern from the United States, an exponentially flattening pattern. Thus, assuming 328 MBTU per person, as shown in the U.S. energy report for 2003, and a present worldwide use of 400 QBTU, we can extrapolate for the near term (2150) a world population of twelve billion people.[8] Energy consumption worldwide would be $12 \times 10^9 \times 328 \times 10^6 = 4 \times 10^{19}$ or 40 QuintillionBTU, or 100 times today's consumption; a target easier to meet with nuclear power than the food supply needed to feed 12 billion people.[9] The asymptotic flattening of energy use to a sustainable level (if nuclear management is figured out) is another example of maturing technological elements in an industrially stable society.

CONCLUSIONS FOR LIFE UTILITIES

By using our *mature design and manufacturing technology* skills wisely, we can develop whatever we can describe clearly in a specification that proscribes durability, cost, function, beauty and material reuse. Again, the problem is more one of defining what is good and what is not than whether it can be built. Society must decide what it wants.

What mechanisms do we have to drive these decisions? Direct government controls have often proven to be inefficient, even counterproductive, while free market mechanisms often do not incorporate all of the important factors, such as pollution and resource loss, on a timely basis. One promising possibility might be for government to use legislation that incorporates cost imposing factors that would not otherwise be included. For example, we consider air to be free, because in the past we have been able to use unlimited quantities at no cost. We could subject air and water to a regulatory price mechanism. That is, you can use as much as you are willing to pay for. This is already being done, or is being discussed as part of the Kyoto Protocol's carbon dioxide restrictions. The government issues credits to businesses, each credit allowing a specified quantity of pollutants to be emitted. The credits can be bought, sold, or traded between companies. It is hoped that this approach will produce less bureaucracy and inefficiency than traditional regulatory methods. Natural free-market price mechanisms can be very effective, and operate mostly without need for additional government involvement, except to ensure that all relevant elements (economic costs to society arising from pollution, for example), get factored into the cost and price.

The consumption of energy is a new and significant part of our lives and continues to improve quality of life even in the poorer countries. Wise use of the remaining fossil fuels and logic driven development of nuclear power for the long run should be right up there with food supply and health care as global priorities.

We must proceed carefully with nuclear reactors, but proceed we must.

Reference notes

1. A complete web history of Computer Aided Design can be found at: www.mbinfo.mbdesign.net/CAD-History.htm.
2. This conclusion was derived from United States 2004 Gross Domestic Product tables at: www.bea.gov/bea/dn/home/gdp.htm.
3. A physics friend of mine suggests there really isn't any non-recyclable trash, only items that decay more rapidly than others. The material content of the world is largely unchanged (Thanks Mr. Parker).
4. Pogo's words, from a Walt Kelly poster used on Earth Day, 1970.
5. *Annual Energy Review* 2003, Energy Information Administration of the Dept. Of Energy. www.eia.doe.gov/aer.
6. Quadrillion is defined using the American system, which means it is 10^{18} BTU's. (Europeans define the word quadrillion differently). A BTU is a physics defined amount of heat, approximately the amount of heat released when a kitchen match head burns. An MBTU is a thousand BTU.
7. World Health Organization Report, Estimated Long Term health Effects of the Chernobyl Accident, April 1996, Principal author: Elizabeth Cardis, International Agency for Research on Cancer, Lyon, France, www.uic.com.au/nip22app.htm#cardis.
8. United Nations *Population Report*, 2000, www.un.org/esa/population/publications/sixbillion/sixbilpart1.pdf.
9. Kendall, H.W., and D. Pimentel. *Constraints on the expansion of the global food supply* (Ambio Issue 23, Royal Swedish Academy of Sciences, Stockholm, Sweden 1994), 198-205, www.ambio.kva.se.

CHAPTER 5

Table 5   The Body

CHANGE EVENTS	YEAR	CHANGE AGENT
X-rays	1901	Wilhelm G. Roentgen
Genetics and Inheritance	1902	Mendel/Fleming/De Bries/ Sutton
Xray CAT Scans	1926	Bob Ledley
Beam Magnetic Resonance	1944	Isidor Rabi
DNA ID'ed as Genetic Material	1944	Oswald T. Avery
Nuclear Magnetic Resonance (MRI)	1946	Edward Purcell and Felix Bloch
DNA — double helix structure	1953	James Watson/Francis Crick
Sub-Four Minute Mile	1954	Roger Bannister
Genetic Code — DNA=>RNA=Protein	1961	M. W. Nirenberg/G.Khorana
Real Time Ultrasound Tissue Scanning	1964	Siemens Inc.
Cloned Living Frog	1970	John Gurdon
Recombinant DNA— Restriction Enzyme	1970	H. Smith/K. Wilcox & P. Berg
Imaging-Magnetic Resonance	1973	Paul Lauterbur, UNY
CAT Scanning	1979	Cornack Hounsfield
Human Genome Project	1991	J. Craig Venter-Celera/ NIH

Table 6   Health

CHANGE EVENTS	YEAR	CHANGE AGENT
White Rice vs Brown (health factor)	1901	Christaan Eijkman / Gerrit Grijns
Wassermann Test for Syphilis	1906	August von Wassermann
Importance of Vitamins	1906	W. Fletcher / F. Hopkins
Vitamin chemistry — an Amine	1911	C.Funk
First Vitamin (A) Isolated	1913	DavisandMcCollum-U.Wisc./ OsbornandMendel-Yale
Blood Bank	1914	Albert Hustin
Chlorination of Water (purification)	1915	Abel Wolmand
Forty million influenza deaths	1918	Swine (Spanish) Flu
Insulin for Habilitation of Diabetics	1922	Banting,Best,McCleod,Collip, U.of Toronto
Penicillin discovered	1928	Alexander Fleming
External Pacemaker Prototype	1932	A.S. Hyman

Table 6   Health (continued)

CHANGE EVENTS	YEAR	CHANGE AGENT
First Synthetic Vitamin — Vitamin C	1933	Szent-Gyorgyi/Waugh,King/ Reichistein/Haworth
Sulfa Drugs	1935	Gerhard Domagk-IG Farben/ P. Gelmo
Penicillin produced	1939	Florey & Chain-Oxford,UK/ Morey/Kane-Pfizer
Pap Smear Test	1943	G. Papanicolaou, H.Traut-Columbia
Cortisone	1944	Julian at DePau./ Sarette at Merck/Kendall
External Heart Pacemaker	1951	J. A. Hopps-Banting Institute
External Portable Resuscitator	1952	P. Zoll — Beth Israel Hospital
Salk Vaccine for Polio	1952	Jonas Salk
Kidney Transplant	1954	Joseph E. Murray
Tetracycline — (broad antibiotic)	1955	Lloyd Conover
Pacemaker (w/ext. battery-heart wires)	1955	C. W. Lillehei and E. Bakken
Interferon (the atom of biology)	1957	A. Isaacs and J. Lindenmann
Implantable Pacemaker	1958	W. Greatbatch & W. Chardack
Medicare Act of 1964	1964	Lyndon Baines Johnson
First Artificial Heart (short success)	1966	Michael Ellis De Bakey
Heart Transplant	1967	Christian Barnard
Artificial Heart (partial success)	1969	R. S. Jarvik and P. Winchell
AIDS Cocktail (to slow the effects)	1996	NIH — many contributors

"DNA is included in this summary more because we became humbly aware of its future impact on us than for its million person affect in the 20th Century."

HEALTH AND UNDERSTANDING THE BODY

T he increasing knowledge of body science and subsequent growth in health sector businesses changed our understanding of and our social behavior toward human incapacity. These two lists most simply describe the major events and discoveries involved in changing attitudes toward our physical wellbeing. The summary below describes this fantastic leap forward from the previous thousand years, a time when we were not able to cure anything that could not heal itself, to the present when many of the most prevalent diseases are curable. Even simple painkillers, such as aspirin, were not invented until the 20th Century though they trace their roots to existing plant-based remedies of earlier centuries.

During the 20th Century we gained mastery over many physical disabilities as the result of:

- The fairly complete understanding of bacterial infection

as well as the etiology, the root cause, of bacterial and viral diseases;

- The development of methods for prevention and cure of infections and some formerly incurable diseases;
- The development of molecule-based biological treatments and cures for new viral diseases, such as AIDS;
- The discovery of DNA and the double helix, leading to ways of modifing naturally evolved and selected DNA, allowing us to eliminate weaknesses in plants and animals more quickly than evolution and natural selection would;
- The development of new body-interior imaging machines that provide extensive and early health diagnostics;
- The performance of increasingly complex surgical (bio-mechanical) interventions, including use of robotic surgical machines;
- The development and installation of electro-mechanical biological replacements and implants for body part repair.

Society refocused its energy away from expenditures on war and toward minimizing disability. We saw the emergence of:

- Health care as a broad and deep profession employing about 20 percent of the working population;
- The societal awareness that we are beginning to understand the fundamentals involved in creating human life;
- The social assumption of health insurance, both private and government-funded, as a way of protecting the unlucky and extending the lifetime of the aged at the expense of the healthy;
- The increased size and emerging social influence of the older populations (average age at death grew from 48 to 76 over the century);
- The immersion of society in the huge economic impact of health care services amounting to a whopping 12% of GDP in the USA.

Take a brief minute to peruse the preceding two lists regarding innovations in body and health knowledge. Think about a time when these possibilities didn't exist. Then consider what remains to be understood and accomplished.

UNDERSTANDING THE BODY

Wilhelm Roentgen, a man whose name is synonymous with the word x-ray in Europe, invented the x-ray vacuum-tube. This gave us the ability to photograph a living human skeleton using film to record x-rays that had penetrated the skin, organs and bones, leaving a diagnostic rich, variable density image on film.

From x-rays to CAT scans (computer aided tomography, the result of focused moving beams of x-rays whose image pieces are reassembled by a computer to make a final image), to MRI to Ultrasound imaging, we became able to see first the bones and later the organs and brain, soft tissue, tendons, heart valves, all moving in real-time. Finally we were able to observe brain activity which led to a revolution in the science and understanding of our bodies' functions and malfunctions. Seeing inside the body is now an essential element of most second level diagnosis for complicated diseases. Doctors invariably use imaging machines to determine the specific treatment for organ cancers and heart malfunctions.

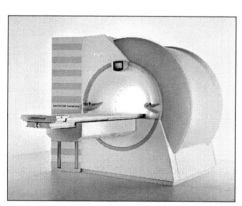

Siemens MRI machine uses a superconducting magnet.

Finally the lid was pried off the developing field of genetic modification and animal cloning by a sequence of discoveries. Beginning in the previous century with Mendel's sweet pea inheritance of traits, then followed in the 20th Century by the discoveries of chromesomes, DNA

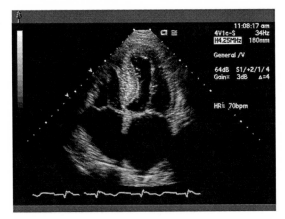

The lower ultrasound image is same heart as above but with open valves (Sequoia 512).

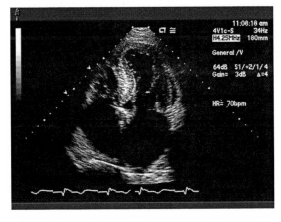

and its double helix structure, we entered the secret world that controls the heredity of all living things. DNA is included in this summary more because we became humbly aware of its future impact on us than for its million person affect in the 20th Century. Modified DNA has not yet affected millions of people's lives. Mapping the human genome or DNA chain was a giant step in understanding our molecular heredity. That mapping was completed shortly after the beginning of the 21st Century.

Synchronous with these inventions and discoveries that helped us understand our body and its function, were the discoveries listed in table six that enabled us, for the first time, to make major repairs to ourselves.

HEALTH

Without forgetting that cures for disease began with a smallpox vaccination in 1771, the 20th Century had a huge impact in all areas of human health care. Smallpox was, in fact, eradicated in the most recent century. For now we look broadly at the magnitude of influence the two lists (Table 5 and 6) describe in our journey to improving and controlling human health.

The century's health diagnostic discoveries began by developing a test for the sexually transmitted scourge of previous centuries — syphilis. Twenty-two years after diagnosing the disease, investigators found a cure for syphilis and fifty years after that a new sexually transmitted disease, AIDS, emerged to become another pandemic. Tests for AIDS exposure became the norm in the 1980s, and by century's end a mitigation had been found that slowed the disease's progress. Though it was not a cure, it allowed near normal life for decades. All of these medically invented tests and treatments became important to a burgeoning sexual freedom initiated as a consequence of the availability of the pill. And, as in days past, sexual freedom and its diseases were also fuel for those ready to legislate morality as discussed in the chapter on social change.

Two of the most beneficial discoveries were cures for bacterial infections. The use of penicillin and sulfa drugs reduced the consequences of infections caused by accident and war and made invasive surgery safer.

The surgical removal and replacement of major organs (kidneys and hearts) to save lives also became more feasible. Later, new antibiotics provided stronger defenses against almost all bacterial infections. Mechanical medicine gained significant ground when surgeons began to use body invasive techniques and installable devices to cure and prevent sickness and to repair damage. Mechanical medicine replaced ligaments in the knees with cadaver ligaments, removed malfunctioning organs and obstructing tumors. They also replaced failed body parts such as hearing systems, heart valves and hip joints with engineered components. They cleared blockages from arteries and veins and repaired broken bones with screws and wire. Surgeons provided numerous repairs of body parts by the end of the century. Only brain repair seemed still out of reach.

In the coming century there will certainly be new prosthetic devices such as artificial eyes, robotic digits, artificial pancreases and other mechanical adjuncts to our bionic repair catalog.

The 20th Century provided a series of palliative (symptom reducing) medications and the beginnings of electronic prosthetics as well. Synthesis of the hormone insulin allowed hundreds of thousands of people with diabetic deficiencies to reclaim normal lives. The Salk and later Sabine vaccines prevented the devastating disabilities of the polio virus. Electronic pacemakers extended the life span of patients with heart timing problems. Heart valves became replaceable. Hearing implants allowed bypassing the entire mechanical hearing apparatus — the ear itself, the drum, the three bones, the cochlear channel and the hair cells.

During the century the mean age at death went from 48 in 1900 to 76 in 2000. This is a 50% increase in useful lifespan. Some of this increase, however, was due to reduced child mortality, though significant gains were made in raw survival as well. (In early 1900-1902 over 20 percent of children died before the age of 6.)[1]

By mid-century, government-funded medical care had become national policy in many countries, providing a broad social blanket that protected the disabled and the aged. In the United States President Lyndon Johnson enacted and signed the Medicare Act of 1964. In the 1950s and '60s many industrialized countries either preceded the U.S. in government coverage or followed suit. Also in that same time frame, life insurance policies, which had begun in the early 1900s, moved to support health insurance by collecting yearly premiums from employers or employees to protect them from disastrous losses of a health crisis.

This tendency toward nationalization of health care (especially in Europe and the communist countries) emerged largely during the 20th Century, but was more an evolving program than a breakthrough of social conscience. The changes in biological and chemical understanding of the human body and the associated medical treatments were more responsible for the move toward national health than a reluctant populace or government. By the end of the century, health services were the largest single expenditure in the Gross Domestic Product of the United States. A recent study of the impact of nationalized health in the United Kingdom indicated that the normal lives of its populace had not been extended much beyond the average age of seventy during the last twenty years of the century. There is some indication that the cost-benefit of health insurance vs. care are not keeping pace with one another. There may be a looming crisis to be dealt with in the 21st Century between cost and value of health care for those over the age of seventy. All nations find themselves in the throes of evaluating society's ability to finance technological health care to its fullest.

On the other hand entitlement for the disabled has gathered more support recently, especially in the United States. Present day workplaces must make accommodation for disabilities that are not contrary to the work at hand. A couple of examples of such disability barrier reductions are: the provision of electronic amplification for the hard of hearing (when the work requires being on the telephone) and access ramps and an oversized rest room for those in wheelchairs.

Toward the end of the century a blending of Eastern and Western medicines and styles of medicine emerged. The Eastern (Asian) medicine had evolved over many thousands of years of empirical evaluation. In China as well as India an entire body of knowledge existed with regard to herbal-medicine. This change had not emerged enough to be considered part of the multi-millions of people paradigm suggested for the lists in this study, but was clearly a trend late in the century.

Thousand-year impact? Let us try to restate the key impact areas listed at the beginning of this chapter and reevaluate them in a reviewer's light:

Clinical
1. We came to understand bacteria, viruses and infection.
2. We can prevent and cure diseases by vaccination and antibiotics.
3. Molecular-biology weighed in to provide chemical investigative means and cures for diseases.
4. With the discovery of DNA we are gaining control of the fundamentals of life.
5. Imaging machines provide powerful diagnostic and investigative techniques to extend our knowledge of human bodies.
6. We mastered complex surgical interventions.
7. We created several electromechanical implants for body part repair.

Social
1. A major increased in size and social influence of the older population emerged.
2. The economic impact of health care became large: 12 percent of GDP in the USA with a similar impact in other industrialized countries.

With the thousand year view, the changes from the 20th Century suggest, primarily, a significant increase in spending on medical diagnostics and treatment from previous millennia as well as a massive redistribution of labor to the profession of health care worker. The century contributed epoch defining increases to our understanding of the human body, achieving, near the end, an ability to create new body parts and to alter the chromosomes of unborn babies. Absolutely no progress was made, however, in dispassionate assessment of the economic value of the decisions involved in making these changes and enhancements, nor of the limits of cost that make sense for society.

Very significant inroads were made against viral and bacterial premature death, ending centuries of capricious and now unnecessary grief. Individually, millions of people were spared. As illustrated in the figure below, survival and life span provide each one of us with personal value, achieved from these advances in knowledge.

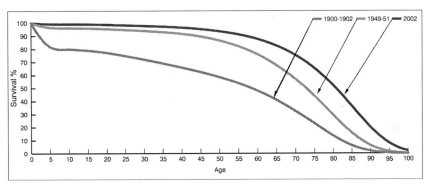

Survival curves illustrate the increase in length of life and reduced infant mortality over the span of the 20th Century.

Reference notes
1. *National Vital Statistics Reports*, Vol. 53, No. 6, Figure 3 November 11, 2004, www.cdc.gov/nchs.

CHAPTER 6

Table 7   Fine Art, Music and Dance (also see Chapter 1 for Music)

CHANGE EVENTS	YEAR	CHANGE AGENT
"Serpentine Dance"	1900	Loie Fuller
"Dance of the Future"	1903	Isadora Duncan
Modern Art (plus Hundreds of artists)	1907	Pablo Picasso
Ragtime Respectability — Carnegie Hall	1912	James R. Europe, Clef Club Band
New York Armory Show of Modern Art	1913	W. Kuhn, A. B. Davies
Classical Guitar	1920	Andres Segovia
12-tone Serial Technique	1921	Arnold Schoenberg
Creole Jazz Band	1923	King Oliver, Jelly Roll Morton
Father of Blues	1927	Son House
Classical Music — Musical Theater	1928	George Gershwin
"Lamentation" "Appalachian Spring"('42)	1930	Martha Graham ,A. Copland
Stereophonic Sound (dual sources)	1931	A.D. English
Large Structure/Sculpture Mobile	1932	Alexander Calder
Jazz Solo Guitar	1930s	Charlie Christian / Django Reinhardt
American Folk Music	1937	Woody Guthrie
Major Builder of The Blues	1936	Robert Johnson
Middle America Realism Painting	1937	Andrew Wyeth
Jazz Vocal Recordings	1938	Ella Fitzgerald
Jazz/Blues	1943	Muddy Watters
Classical Ballet	1948	George Balanchine
The LP Vinyl Record	1948	Peter Goldmark
Popular American Folk Music	1949	Pete Seeger
Rock and Roll	1954	Elvis Presley
Opera	1965	Luciano Pavarotti
Tonality Influence on Rock — Beatles	1968	Ravi Shankar
Black and White Photography as Art	1902	Alfred Stieglitz
Movie Comedy	1915	Charlie Chaplin
Movies: MGM	1917	Louis Meyer

Table 8 Recording of Life Past, Present and Future

CHANGE EVENTS	YEAR	CHANGE AGENT
Color Film	1928	George Eastman
American Ideal Family Paintings	1930	Norman Rockwell
Polaroid Film	1932	Edwin H. Land
Brave New World	1932	Aldous Huxley
The Rise and Fall of Cultures	1934	Arnold J. Toynbee
Plays of American Severity	1936	Eugene O'Niell
Television	1939	David Sarnoff
Stage Musicals	1943	Oscar & Hammerstein
"All in the Family" real people	1971	Norman Lear
"Hill Street Blues" real TV drama	1981	Steven Bocho

Table 9 Poetry and Literature

CHANGE EVENTS	YEAR	CHANGE AGENT
Transition Poetry	1900	William Butler Yeats
Offset Printing	1904	Ira Rubel
The Jungle	1906	Upton Sinclair
The Hollow Men	1925	T.S. Eliot
Animal Farm	1945	George Orwell
East of Eden	1952	John Steinbeck
The Lord of the Flies	1954	William Golding
American Slash Poetry	1960	Anne Sexton
To Kill a Mockingbird	1961	Harper Lee
2001: A Space Odyssey	1968	Arthur C. Clark
A Brief History of Time	1991	Stephen Hawking

Table 10   Humor

CHANGE EVENTS	YEAR	CHANGE AGENT
Americanization of English Humor	1925	James Thurber
New Means of Conveying Comedy	1926	Buster Keaton and Harold Lloyd
Standup TV Comedy	1930	Marx Brothers
Comedy Slapstick	1934	Three Stooges

Table 11   Fantasy

CHANGE EVENTS	YEAR	CHANGE AGENT
Disneyland	1955	Walt Disney

"...art is a summary of what is happening in the culture overall."

THE ARTS

This chapter reviews the significant events and contributions involving the arts over the span of the 20th Century, discussing and classifying them for their relevance from a thousand-year point of view. To present a complete description of the correct names and magnitudes of accomplishment in the arts and to discuss each in detail would require another book, a book written by an art historian. My intention is to make readers aware of the 20th Century's bigger picture, and use examples that are the most strikingly representative to reveal the underlying character of the century. The spreadsheets of this book encourage you to make changes and additions or deletions, as you see fit.

Let's first take a look at the listings of different sub-categories of art to be discussed in this chapter (Architecture has been placed in a later chapter related to the author's sense of appropriate grouping, without implying architecture is not an art).

THE ARTS

Much like our subconscious thoughts, our therapists' opinions and our own 20/20 hindsight, the arts provide us with symbols that show what is going on in the bigger picture of our lives while we find ourselves immersed in its details. As expressed in Kenneth Clark's public television series, *Civilization*, art is a summary of what is happening in the culture overall. That includes not only the art on display in museums but also art displayed through the form of our buildings, monuments, performances, and also that which has collected in our archives.

The 20th Century artists developed and adopted new ways of expressing universal feelings and realities. The forms, as well as the subject matter reflected social, cultural, technological, and political characteristics of the times. They not only interpreted, but greatly influenced 20th Century life, becoming intertwined to the extent that it was not always possible to separate cause from effect. The arts are at once very international and circumscribed by continents and cultures. This author's review is confined very much to Western art. This narrative inherently lacks reaching into Asia and the Middle East as a result of the author's own limits in those cultures. It is not the limit of exposure to the many continents' outpourings so much as how the 20th Century component of Asian and Middle Eastern art was altered and whether it changed in far reaching or minor ways. One must almost be raised in a culture or become a traveling specialist to make that judgment.

Because jazz and rock and roll emerged in the Western culture and spread worldwide in popularity they will be internationally representative, but the photography, dance, sitar music and literature examples will not be representative of non-Western sources.

The subcategories chosen for the arts were arrived at by the author after preparing a master list of events, discoveries and inventions affecting multiple millions of people. These categories were then compared to similar groupings in other published cross-cultural histories[1] and in some cases renamed or re-grouped. Finally some were kept because, to record life past, present, and future seemed to me a more relevant approach for that form of art.

I found the following to be my conclusions about the significant impacts on us as a result of 20th Century art's development:

- Jazz, rock and roll and movies, became a totally new vehicle for worldwide understanding and communication, beyond language, and moved us toward global unity.
- Classical music did not change in any significant way. It did not find its way into abstraction as classical fine art did.
- Movies allowed us to experience many lives within our one life.
- Movies and fantasy project us into non-existent worlds. Our visual experience was enriched beyond the most fantastic of our nighttime dreams.
- Movies and TV created severe social isolation in the developed nations.
- Abstract fine art was created. It borrowed new materials and invented new forms for its expression. Its impact on the future of representational art was large.
- Photographic art emerged as a new representational art form.
- Patron support of fine art became less the prime source of funding and was replaced by consumer and government-financed art.
- Dance was completely altered by a new freedom of movement.
- Literature reached its greatest level of achievement in the 20th Century, its golden age.
- Poetry's content and impact remained unchanged, but its form was irrevocably altered and it became much more an every-man form of expression.

Let's discuss the multiple lists of accomplishments in the arts to see how I arrived at these conclusions about the century's artistic output with thousand-year consequences.

FINE ART, MUSIC AND DANCE

Fine Art Breaks Away

What we call fine art had moved quite forcefully into the modern era with Picasso and Braque's cubism first, and thereafter into all the forms of modern painting including the paint throwing Pollock, the paper cutouts of Matisse and a line of notables listed more fully in other studies focused on modern art. The secularization of art, its movement away from religious themes, occurred most completely in previous centuries and is only mentioned here to explain its absence. What I do emphasize, relevant to the thousand-year impact focus of this book, are these three points:

1. Non-representational art — Abstract art

There is a known linkage between technology's creation of photography in the 20th Century and the resultant need for fine art, painterly art. That linkage suggests we reassess art's value by finding new ways, other than perfect image reproduction, of capturing life and meaning in images by a move toward abstraction (non-realism) in art. In addition, our industrial world, filled with industrial buildings and machines, added to the artist's visual space by providing images never before painted, entreating artists to move their imagery through the many artistic spaces previously traveled of landscapes, sculpture, portraiture, mythology and juxtaposition.

2. Alternate locations, materials and dynamics for display

The departure from the representational provided new freedom to display in alternate spaces (Christo's *Running Fence*), to use new material(s) for media (plastic, light pipes, video) and to change the immobility of the art itself (Calder's mobiles).

3. The inner landscape

Lastly, though earlier fine art had, as far back as Giotto, revealed inner agonies and feelings, the emergence of Freud's ideas of the psychological journey inward to the subconscious led 20th Century artists to take responsibility for revealing images of that inner landscape. The art of feeling and psychological chaos is new in the 20th Century, though in some ways it may be linked to subconscious forces present in Hieronymus Bosch's *Garden of Earthly Delights*.

Quoting from a modern San Diego gallery's description of abstract art: ". . . non-representational art is still the most probing, powerful, and effective way of going about the rendering of modern life. Art is not, however, solely about preparing the history that will be written about us. Its elevation of a single moment is perhaps just as important as its exploration of broader matter. Art has the ability to ease suffering through empathy as much as it has the ability to enrage and disturb through provocation."[2] Art's abstract century emerged for the United States at the 1913 Armory Show in New York City, where many of the images that would dominate the century hung waiting for judgment.[3] The support of art by rich patrons, especially the government and the church during the Renaissance, deteriorated over subsequent centuries and finally financed only the buildings housing the artists in the 20th Century. Private purchases by individuals worldwide, as well as thousands of gallery sales, became central to artists' livelihood.

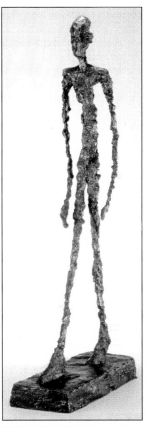

Walking Man II, Giacometti

I pause in this writing to think, "how can I, from this time forward, imagine human form without recalling Giacometti's sculptures re-defining the human form emotionally, nearly equaling Michelangelo's Pieta in Rome?" Other examples (happily you'll have to look some of these up) that come to mind are :

- *Nude descending a staircase* by DuChamps, movement within the unmoving.
- The floating angels of Chagall competing with Orthodox iconography of the 11th Century and the scenes of Revelation from Bosch's 15th Century.
- The timeless clock-formed faces of Edward Keinholz' *Beanery,* a small walk-in drinking-bar sculpture in Amsterdams Stedjik's museum.
- Nam June Paik's televisions and their video streams pressing conscious and sub-conscious emotional reverberation in *Global Encoder.*
- Tansey's Christian allegory, *Doubting Thomas,* where a man kneeling on a narrow cliff road places his hand in the broken open crack in the roadbed, while his wife looks on through the driver's

open door, a gentle mimicry of the Bible's Thomas touching Jesus' side to establish that he is real.

- Picasso's *Guernica*: A cry of sorrow and rage against human suffering in Spain's civil war. Has any photograph from ravaged Vietnam equaled it? Though two images do come to mind (First the naked, excruciatingly thin 9-year-old girl running down a muddy street holding her hands to her head in fear and loss, and second the general executing a man in the streets of Saigon using a pistol to his head). It is amazing that Picasso's art on canvas can so haunt us. The fact that the Vietnam photos existed and were seen by so many people that the course of the war may have been altered, shows how cultural, technological, and geopolitical events merge toward action.

Pablo Picasso's Guernica. A depiction reflecting horrible suffering during the Spanish Civil War after the 1937 Nazi assisted bombing of Guernica in Basque, Spain. The painting size is four humans wide and two humans high.

These images of modern abstract art are so revolutionary as to have altered our mind's access to reality and brought new forms and expressions of art to those returning to, or remaining in, the representational. It is through this departure from representation and return to it that modern art has most enabled us to see again what is essential. . . to the heart, core, the elemental-quark of what is human.

During the 20th Century representational (classical) fine art evolved, but did not make thousand-year significant changes. The century added many new, high quality representational pieces to collections, but did not change representation. Though rated mundane in artistry by many critics, Norman Rockwell, Andrew Wyeth and other American-life artists of that time, are likely to become the Breughels of the 20th Century, painting classic and lasting Americana.

Jazz and Rock and Roll — A New International Language

The early origins of rock and roll are murky at best. My list will not attempt to be a definitive work on who deserves the most prominent places under the musical sun. As this story of worldwide music suggests, however, it is again the large-scale involvement of human beings in the evolution of jazz along its branches that is more important than the naming of the individuals themselves.

Combinations of Christian hymns, modified by slaves into gospel and soul music, combined with the fast paced energy of Appalachian hills' country violin (probably rooted in European gypsy and Cossack traditions) coalesced into this new form of music and thrust it onto the world stage. Son House and Robert Johnson are two early jazz and blues names deserving visibility. The vinyl record industry bulges with the recordings of early and later jazz styles and its revivals in the mid-century. And with the fading of '60s era rock and roll during the last decade of the century, jazz reasserted itself in many listeners' repertoires as the primary form of music.

During the depression (1929-1939) the roots of rock and roll began to grow, starting with contributions from men like Robert Johnson, as jazz had a decade earlier emerged from African-American roots in the South. By the mid '50s, rock and roll had entered every American teenager's heart with Elvis Presley's releases of *Jail House Rock, Hound Dog* and *Love Me Tender*. During subsequent decades rock and roll firmed up its grip on the teenagers emerging from the post World War II baby boom. Rock and roll expanded its style and audience immeasurably on the coattails of British mainstream music started by the Beatles. It spread across all continents and languages. It spread behind the Iron Curtain in spite of government controls, into the USSR and continued on into China. In the end, rock and roll broke open barriers to communication between East and West that had existed for centuries in language and culture. To this day, China plays English jazz and rock and roll at restaurants for the locals.

Folk music in the era of the '60s embodied that central core of all folk music, the provision of pathos buoying a local social discord that rung true to a broad base of people and creating in them a sense of urgent belief that spurred them onto the barriers of release. Folk music provided a voice in '60s America for the elimination of racism and a pressure against the Vietnam war.

This new worldwide communication via music was as central in the effort toward world peace and nuclear disarmament as were the diplomats specifically assigned to that task. On the other hand these same songs and movies have incited violent reactions in cultures where a song's message is contrary to local mores or beliefs, as it has been in the conservative sects of Islam as well as Christianity.

The simultaneous emergence of low cost and high quality music using electronic reproduction made a worldwide audience possible. Worldwide growth was stimulated by the availability of audio electronics, records and later CDs through worldwide distribution. Bands arrived by airplane to initiate young audiences. Our era of mobility and communication was upon us.

The Beatles and the entire '60s rock and roll scene took youth culture into its grip in a way never witnessed in the annals of human behavior. Folk music especially captured young peoples' hearts and pulled them into a spirited liberalism of sexual behavior that leapfrogged even late post-Victorian limits. This musical abandon, combined with the birth control revolution, became another source of social pressure attempting to reset the limits of cultural norms for sensuality.

The guitar became ubiquitous, mostly through its central role in the bands performing rock and roll, folk and country-western music. Other forms of the guitar such as the electric sound-pickup, non-acoustic guitar gained musical dominance and drove the guitar into prominence through rock and roll's outpouring.

The words "classical guitar" were invented for wide audiences outside of Spain, with the emergence of Andres Segovia's performances of Fernandez Sor's 19th Century compositions. These classical pieces, combined with Segovia's virtuosity (and later his own compositions) became the definition of classical guitar. Meanwhile, in India, Ravi Shankar spread the Sitar's new tonality to the west via releases into rock and roll through the Beatles. Both of these uses of stringed, lap-operated instruments were elevated and classical. The guitar of rock and roll was savage and in that savagery released a pent up energy in its listeners. It is the dominant instrument of modern music followed closely by the electronic piano called the keyboard.

The Century's Backwater of Music Innovation: Classical and Opera

Classical music's most recent century credited with inducing thousand-year changes to composition was the 18th, the century of Johann Sebastian Bach and Wolfgang Amadeus Mozart.

Thereafter small changes continued along similar lines to those of abstract art to fine art, and by the end of the 19th century various harmonic extensions to the "common practice" of western composition had reached the point where shifts in tonal centers (key) were so rapid and far reaching that there were no longer any clearly defined tonal centers to many compositions. In addition, simple harmonic cadences with clear key centers gave way to more unresolved (dissonant) preparations for cadences that never happened. Schoenberg took this trend to an extreme with "Transfigured Night" and other early pieces. These are dissonant pieces only in the sense that there are few clear cadences in any key, and there is a consistently dissonant texture to the intervals.

Dissonance became a grand experiment, as abstraction and non-representation did for fine art. Classical music attempted several moves toward abstraction, but did not move far enough from classical forms with their experiments, even if we include rock's heavy metal and synthesized music, to allow classical music to make more than modest evolutionary steps year by year.

My sense is that classical music has not made one of the thousand-year durable changes during the 20th Century with the single exception of Schoenberg's 12-tone serial composition technique. Schoenberg altered modern composition by establishing the 12-tone serial technique as an important organizational device. Serialism evolved because of a concern that without a systematic means of organizing tones, music would degenerate into an undisciplined and chaotic exercise. Some think modern music already has.

In the current era (21st Century) the richness of the listening environment (thanks to the incredible inter-connectedness offered by modern technology) has grown exponentially. This has slowed down the process of integration of the various elements of the world's music. Instead there are almost too many styles, perhaps all waiting for another Bach to integrate them.

Opera saw minimal change as well. There was a reemergence of powerful male voices, though again they are most notable in the way of the sports hero — as outstanding personalities, not in the way of changing opera itself, and are therefore not included here though they were contained in the lists.

There were many continuing contributors to classical composition, none were part of the discontinuous, classical-altering events this book has chosen to examine. Assonance was embraced, but, unlike the art world we discuss below, this abstraction of music did not lead to new, widespread, ubiquitous awareness of music's reinvention.

Dance

Dramatic innovations in the world of dance gave fluidity driven by Art Nouveau images as illustrated in the gilded Art Nouveau statues of Loie Fuller. Later and more completely, Isadora Duncan and Martha Graham choreographed this fluidity. The dancers brought Art Nouveau to dance with graceful arcs and flowing scarves, with the abandonment of formal costumes in favor of full revelation of emotional range through movement. Dance, though not normally affecting multiple millions of people, and considering as well the similarities in new Western dance to Arabian and Indian forms, added a changed and more fluid flow of movement and informality to the Western reservoir of visual and musical art in the 20th Century. Arabian, Indian and Chinese dance had contained fluid motion for many centuries.

Loie Fuller brought the Arabian dance of veils into western culture. Isadora Duncan brought new ideas about the fluidity of the human form to dance, and broke away from the formally costumed processes of classical

Raoul Larche's early 1900's sculpture depicting Loie Fuller dancing with veils.

ballet. Martha Graham's studio led the completion of these transitional changes to modern dance. Quoting from the public broadcasting system's web page, "Graham believed that through spastic movements, tremblings, and falls she could express emotional and spiritual themes ignored by other dance. She desired to evoke strong emotions, and achieved these visceral responses through the repetition of explicitly sexual and violently disjunctive movements."[4]

These were major changes in dance, akin in magnitude to the Modern Art revolution for fine art in terms of scope and impact.

RECORDING OF LIFE PAST, PRESENT AND FUTURE: FILM MACHINES

As improved methods for making accurate records of life and life stories on both film and magnetic media developed and became dominant during the 20th century, those records emerged more and more as a form of cultural continuity and education. They embodied a vital new cure for the consequences of Santayana's famous warning, *"Those who cannot remember the past are condemned to repeat it."*[5] Movies became a means of incorporating the experience of many different lives and cultures into our own lives.

Movies expanded our individual awareness beyond its narrow confines and extended learning and emotion from our past into our descendants' futures. Movie animation broke new ground, creating imagery so true to life as to defy our present realities. Such movies as *Jurassic Park* and *Star Wars* delivered fantastic worlds of apparent visual truth.

The section of the arts list provides examples of life's records given to us by some of the best chroniclers of the time in their medium. Stieglitz' photographs began the era of the photograph as Art. The novel *Brave New World* revealed an ominous future of social weakness caused by the over-use of drug-induced happiness. Charlie Chaplin's mix of humor and frustration, in silent movie form, liberated us to see ourselves with humor in the midst of our struggle. Eugene O'Neil's tangled web of familial past and present transgressions dragged all who entered into its downward emotional spiral, warning us off (if that is possible). Toynbee's popular cultural history provided glints of historical perspective to assist us in internalizing movie and television lessons, providing vision and thus modifying our behaviors enough to help us make progress in improving the human family.

I have focused on the positive worldliness of visual media. There are also negatives. There was increasing loss of local social intimacy, loss of the neighborhood connectedness in cities and towns, that easy intimacy has been supplanted by the magnetic emotional pull of televised and movie entertainments. Examples of disappearing social behaviors are the missing summer discussions between city neighbors on the brick townhouse porches of Philadelphia in the 1950s, or men gathering in the plazas of Italian hill towns in Tuscany at sunset, or summer strolls along the grassy boulevard promenades of St. Petersburg in 1930s Russia. These losses of local intimacy go some distance in explaining the causes of alienation in today's society. The 20th Century, like many of the centuries before it, presented serious problems relating to the evolution of its societal and technological underpinnings, providing new homework assignments to our emerging generation of leaders.

The consequences of this greater isolation are not easily quantified, and will not be studied in this book. One impediment to its summarization is that the consequences are still caught in the long time-constant of culture change mentioned in the first chapter. What can be said now is that we have lost the close-knit, local, social interaction of a neighborhood, caused as much by automobile mobility as by the diversions of CD, TV and movies.

It would also be difficult to compare anti-social behavior over the centuries, as there is limited data from earlier centuries to compare to today's crazed gunman in a Texas clock tower. In view of the fact that technology has increased the ability of one or two people to wreak havoc on the many, using explosives, guns and computers, to what former century abuses would we compare these events; to small town abuse of individuals by those in power, to enslavement by local tyrants in feudal Europe? Much of such abuse was not as openly visible or recorded in earlier centuries due to limitations in gathering and distribution of news.

Our immersion in television does, at times, mimic Huxley's drug-induced absence from reality, suggested as a disease in his book, *Brave*

New World, but of course it also provides the replay of life stories in everyone's living-room, making us conscious of the past to assist us in modifying our behavior. From that consciousness hope springs that action will be taken to avoid those behaviors. *Hill Street Blues* and *All In The Family* provided two early examples portraying people who made poor choices in their lives. These same television shows could, of course, also be used to portray all aspects of life and might foster poor behavior by glamorizing it.

For the moment I have simply added questions to the sociological mix of the century. This particular conundrum; "does the isolation of visual and musical entertainment do more harm than good?" will not be answered here.

Poetry and Literature

Literature and poetry give us a clearer picture of our collective selves. The release of *The Lord of the Flies* led us to question our presumption of youth's innocence in its simplest interpretation. It challenged our view of social interaction at its most unfettered level in revealing what our young might do without the staying hand of civilization.

The other choices on this list illustrate, from bottom to top, the following clarifications of view:

To Kill a Mockingbird rebuilds fundamental values of tolerance of differences; an influential white lawyer in a small American town defends a black man accused of rape, while a mentally slow next-door neighbor becomes a life saving hero to the lawyer's children.

2001, A Space Odyssey envisions a next step in our evolution, coming to us by external intervention (the monolith representing God) as well as self-discovery (science). After the monolith appears and alters the ape's thinking (intervention again), he uses a bone (tool) to dominate other animals. In this movie, however, before we can make that apelike developmental leap, our own thinking machine, the computer HAL, begins to make decisions for its human shipmates, eventually killing all but one of them (subtract one letter each from IBM and you get HAL).

East of Eden asks the timeless question about choosing a life for good or for evil. It answers that we have a choice and must make a choice and continue to make it every day of our lives emphasizing again that our fate is not predetermined.

Animal Farm suggests the flaws of power when the dominance of intelligence in one animal (the pigs) over other animals allows them to take over a farm only to become abusive and dictatorial to all other animals.

And for Sinclair's *The Jungle*, there are the unintended consequences. After the publication of his book the U.S. Food and Drug Administration was created in order to gain control of poor meat packing and inspection standards he described. Sinclair described unclean conditions within the Chicago meat packing industry. Although it did not inspire the socialist revolution which he invisioned, it did provide the government with a tool to improve working conditions and reduced the populace's exposure to health risks. As mentioned previously in Chapter 2, the improvements in working conditions enforced by the FDA further reduced the lure of Communism to the worker.

In his book, *A Brief History of Time*, Stephen Hawking became the first widely read physicist. From his prestigious Cambridge perch Hawking chose to write about post-Einstinien physics regarding the origins of the universe and the big bang theory. The newly non-religious, searching for a new Myth, are making this type of book a bestseller.

The three poets chosen, Yeats, Eliot and Sexton, introduced us to non-rhyming poems. Thereafter, late in the century, a new era of poetry emerged in the United States, wher poets descended on thousands of coffee houses with self-published poetry chapbooks. This poetry was at once more visceral, though still as full of feeling as any from earlier centuries, less formal and shorter (one-page images). The names of Yeats, Frost, Pound, Cummings, Eliot and Collins ring in our ears as the most well known; many emerged to be powerful emotional forces of the 20th Century.

These stories provide major openings in our visual-verbal self-awareness, portending significant change in our civilization's litri-cultural record. These 20th Century stories describe a world millions of miles from those of Charles Dickens, where life was circumscribed by hardship and stern schoolmasters.

Humor

Umberto Ecco's book, *The Name of the Rose*, states that without humor and the ability to laugh at ourselves we lose an essential ingredient of truth. Thurber, Keaton, the Marx Brothers, and Chaplin and Lloyd provided us with video and written reminders of our ability at self-ridicule, a skill sorely needed to augment our generation's overly serious and focused behaviors. Humor also provides coping schemes that we can imagine for ourselves; Keaton's quiet calm, Chaplin's appeal for help from others and Lloyd's acrobatic cliff hanging suspense and escape from the bad guy by a hair's breadth — precursor to the modern action movie.

Thurber makes extreme fun of married relationships in ways now prominent in television sitcoms.

And Keaton gave us the silent movies with extremes of situation crisis management. His skill as a director and scene maker as well as his own acting in silent movie comedy, made him a lasting monument to life's chaos and image creation. He remained calm, steady and durable through the chaos around him in his films.

Fantasy

The 20th Century's continuing evolution of movies, film, and digital video enabled unprecedented creation of wide ranging imagination-generated video images, i.e. reality-free imaging. Walt Disney's cartoons moved audiences to the very edges of possible reality. Often one cannot tell fabrication from reality. Engineering and new materials also enable sets, robots, and entertainment to reach new heights of accuracy in fantasy, from the undersea ride in a submarine to a ride into the past with dinosaurs. The portrayal of near human robots in movies like *Blade Runner* and *AI (Artificial Intelligence 2001)* make the future imaginable and facilitate its emergence. Walt Disney spawned this imagination-animation and brought it to fruition with his classic 20th Century epic story retellings of Cinderella and Snow White on screen. Later in the century companies like Pixar and Industrial Light & Magic brought a blend of mechanical models and computer generated cartoon images to us, providing nearly real journeys into the 200,000 year past of *Jurassic Park* and the 4000 year future of *Star Wars*.

THE ARTS — CONCLUSIONS

Thousand-year consequences can occur in two ways: by altering the direction of influence established over the past thousand-years, or by producing changes in society with consequences that may last into the next thousand-years.

Jazz, rock and roll and movies became a universal language for all nations, providing a medium to show and feel life as it occurs in different countries, cultures and times. Movie companies make actual verbal or textual translations available, extending their audience to other countries in their local language using subtitles or, for the most popular movies, voices dubbed by actors. Movies are often a caricature of a culture and can misrepresent daily life in that way, but can still provide a far greater awareness of our common cross cultural emotions and heartaches across our planet. Movies display our worst as well as our most heroic selves. Songs and music of love and sadness also convey these caricatures of life using melody as a means of getting inside us without the need to understand the words.

Movies, rock and roll and jazz communicated across national and cultural borders, an occurance fueled by the availability of inexpensive electronics and magnetic and plastic media (CDs, VCR and cassette tapes, IPODs, DVDs, and movies. Just as we become more one world, we also become more isolated from our neighbors — more is going on in our heads than in our outside world.

For literature and books, the 20th Century was possibly the best ever for both quality and quantity of fiction. It us unlikely that the Internet and video-driven world of the 21st Century will further the levels of quality reached in literature during the last century. This implies that we need to protect the world's literature using modern means of digital storage and by continuing to print quality books. Fiction will decline in coming centuries to the level of the comic book as clearly indicated by the New York Times best seller list of the recent decade.

Many of us still find abstract art inaccessible, its meanings unclear, the intended emotions and reactions cloudy. The essence of what happened for art during the 20th Century, however, was abstraction; the creation of visual representations of feeling, emotion and non-representational visions embodying texture, color fields and distortions of reality whose purpose is to reconstruct our understanding. I suggested, in spite of the limits of its reach that abstraction assists in the growth of all art. Many believe it enlivens our awareness of possible realities. It uses the visual to provide a Fourth of July like show, demonstrating new realities of motion and color for our delight and entertainment. Other abstract artists drive us into altered visions of war where droopy baby blue bombs hang in space and machine-gun ammo-belts become infertile fields. The thousand-year future of abstract art is certainly not assured, but its departure from the path of art in the previous thousand years is as striking as Hieronymus Bosch's vision of the last judgment was to the art of its time. Christo's running fence blends the land of earth and God with the fabrications of man, coloring a horizon, recreating a sunset. An honest understanding of abstract art and its message is not always essential in each of our varied lives for it to change art's paths in our future.

Reference notes

1. See selected bibliography at the end of the book.
2. *Fuze Gallery Philosophy*, Eli Fernald, www.fuzegallery.com.
3. http//xroads.virginia.edu/~MUSEUM/Armory/armoryshow.html.
4. www.pbs.org/wnet/americanmasters/database/graham_m.html.
5. George Santayana, *The Life of Reason*, (Scribner's Sons, NY 1905), Vol. 1, 1905.

CHAPTER 7

Table 12 Psychology

CHANGE EVENTS	YEAR	CHANGE AGENT
Interpretation of Dreams	1900	Sigmund Freud
IQ Test	1905	A. Binet and T. Simon
Inside the Mind	1911	Alfred Adler
Beginnings of Behaviorist Theory	1913	John B. Watson
Child Development Phases	1929	Jean Piaget
Skinner Box Modifies Behavior	1938	Burrhus Frederic Skinner
Child Development	1950	Erik Erikson
Power of Positive Thinking	1952	Norman V. Peale
Syntactic Structures	1957	Noam Chomsky
Child/Adult Moral Development	1968	Lawrence Kohlberg
Self-help Book Mania	1990	U.S. boomers born 1946-64

Table 13 Philosophy

CHANGE EVENTS	YEAR	CHANGE AGENT
Interesting Problems	1910	Bertrand Russell
Tractates Logico-Philosophicus	1920	Ludwig Wittgenstein
Philosophy of Existentialism	1943	JP Sartre/Albert Camus
L' Etranger (The Stranger)	1942	Albert Camus
Friendship and Hope Defined	1943	Antoine de Saint Exupéry
Individual Objectivism	1957	Ayn Rand

Table 14 Technical Uncertainty

CHANGE EVENTS	YEAR	CHANGE AGENT
23 Open Math Questions	1900	David Hilbert
Mathematics of Approximation, Computers	1900	Taylor, Hermite, Hamiltonian
Sinking of Titanic	1912	SS Titanic
Math is Uncertain — Gödel's Proof	1931	Kurt Goedel
Philosophy of Science — "Falsifiability"	1934	Karl Popper
Fission Failure —Three Mile Island	1979	Atomic Power Paranoia

Table 15   New Tools

CHANGE EVENTS	YEAR	CHANGE AGENT
Computer Programs/ Artificial Intelligence	1937	Alan M. Turing
Software — Nanosecond Wire	1945	Adm. Grace Hopper
ENIAC Computer	1946	Mauchly/ Eckert /John / J. Presper Jr.
Computer Architecture	1946	John von Neumann
Apple MacIntosh Personal Computer	1984	Steve Jobs, Steve Wozniak

Table 16   New Teaching

CHANGE EVENTS	YEAR	CHANGE AGENT
Montessori Method	1906	Maria Montessori
Historical Novel	1948	James Albert Michener
1984, Brave New World (in 1932)	1949	George Orwell/Aldous Huxley
Dr. Seuss — *Green Eggs and Ham*	1955	Theodore Seuss Geisel
Muppets — Puppet Educators	1969	Jim Henson
Kid's TV	1973	Mr. Rogers/Jim Henson

"The attempts to systematize, catalog and analyze human behavior expanded significantly in the 20th Century."

PSYCHOLOGY, PHILOSOPHY, TECHNICAL UNCERTAINTY, NEW TOOLS, NEW TEACHING

The categories combined in this chapter appear to be an unusual grouping. My logic for grouping them is this: The first two and the last are from the set of the cognitive sciences that form a specialized collection embodying our inner thought processes; the remaining two, technical uncertainty and new-tools, embody the external sciences of darkness and light held out for us to use via these same cognitive sciences on the thought-journeys we find ahead of us.

First my summary of the effects of philosophical, linguistic, psychological and learning-based discoveries, inventions and events of the 20th Century:

- Curative therapies for emotional and mental pathologies were primarily developed during the 20th Century and expanded using chemical treatment to the brain. There is significant evidence of individual improvement, but there are no society-wide measures to date.

• Self-help books for all manner of human disturbance, attainment and activity were published for the first time.

• An understanding of the mathematics and science of language was established.

• An understanding of the fundamental limits of science and mathematics emerged in many studies and papers.

• Mathematics and science also dominated the direction and evolution of philosophical thought in the century.

• The historical novel emerged as the dominant means of extending our ability to remember history to illustrate the successes and mistakes of the past.

• And finally, a vitally important new tool emerged to extend human understanding: the computer and its compute-engine, the microprocessor. This is as important an invention to humankind as earlier fundamental tools were, such as the printing press and gunpowder.

To substantiate these conclusions let's examine first each sub-list separately and assess its larger influence on our century.

HUMAN PSYCHOLOGY

Understanding human behavior and its range and depth has been a subject of thought, experience and literature for centuries. The attempts to systematize, catalog and analyze human behavior expanded significantly in the 20th Century. Sigmund Freud began the century with his book, *Interpretation of Dreams,* which set the cornerstone for much of the work and analysis of individuals' psychological struggles throughout the remainder of the century. Freud's focus at the time pivoted around the side effects of sexual repression as a source of pathology. [1]

Early work on behavioral analysis and modification done by Watson and later extended by Skinner would lead to more practical and scientific techniques for altering human behavior.

Erikson, and later and more popularly Piaget, analyzed child development stages and advanced our understanding of how and when different capabilities in children developed and at what stage those elements of personality became fixed. Their studies and many that followed filled out the processes of human learning and methods to facilitate social learning. The studies of Kohlberg and Bandura in the last half of the century provided indicators for the management of social moral behavior and encouraged methods to lead us toward a unified social morality. The mini-book by Gregory Stock, *The Book of Questions*, asking difficult moral questions was an implemented example of this means of raising one's moral conscience from one Kohlberg-defined level to another — essentially to a broader awareness. Bandura exposed evidence of the large side effects of parent-modeled behavior on children. These were huge additions to our knowledge of the cognitive sciences in the last century. Sadly, in all of this, there is no unifying tie to all of the elements of psychology. Any list of major contributors is woefully inadequate, proof again that modern advances in all fields are made on a broad front. Synthesizing an answer unified enough to teach in K through 12 classes is sorely needed. If high school classes presented a clean summary of this wide-open field of psychology and provided dramatic examples and discussions, such education would be invaluable to teenagers reaching adulthood when their lives gain great social complexity and they seek answers and help.

Recent summaries of the best methods for treatment of psychological problems suggest a combination and integration of the methods developed over the span of the 20th Century, the particular mix of treatments dependent on the doctor's specialty and his/her initial findings with the patient.

The information below illustrates this aspect of mixed treatment and is presented in the final chapter of a recent W.W.Norton textbook, *The Mind, Brain, and Behavior*. This revealing list describes a sequence of treatment options dependent on different personality difficulties.[2]

Basic Principles of Treatment
Psychodynamic Assessment
Uncover unconscious conflicts that give rise to maladaptive behaviors.

Humanistic Assessment
  Help people fulfill potential for growth through self-understanding.
Behavioral Assessment
  Maladaptive behavior is unlearned through conditioning principles.
Cognitive-behavioral Assessment
  Eliminate distorted thoughts that cause maladaptive behaviors.
Group Interaction
  Use groups for social skills and interpersonal learning.
Systems Work
  Improve family interactions as issues arise in larger social contexts.
Biological Treatment
  Directly treat abnormalities in neural and bodily processes.

This list and the list below summarize the blend of approaches developed in the 20th Century that can be used together to treat mental illness. Each clinic uses different treatments depending on their specialty as well as the patient's symptoms.

Art Therapy
Behavioral Therapy
Cognitive Therapy
EMDR — Eye Movement Desensitization Reprocessing
Family/Marital Therapy
Gestalt Therapy
Humanistic Therapy
Play Therapy
Postmodern Therapy
Psychoanalytic Therapy
Group Therapy
Chemical Therapy
Shock Therapy

Many studies document percentages of success, both short term and long term for different therapies, but there is no real indication of overall societal success to measure whether society contains less destructive mental illness than it did before such treatments. This science of psychology has progressed and become a primary tool in helping those with emotional dysfunction in industrialized society. Treatment helps people avoid harmful choices for themselves and those close to them. Its success rate is not fully visible in the literature.

The development of the IQ test for intelligence evaluation and comparison heralded the beginnings of good and bad attempts to assess an individual's brainpower and the causes of differences among people in terms of learning rates and success in life. Recent work by Charles Murray attempts to posit the importance of IQ for success in living.[3]

The work of Piaget in child psychology has been very informative and influential in enhancing our understanding of age related learning, but so far our ability to avoid grown children's maladaptations to life does not seem to have improved. According to Erikson, Kohlberg and others, during childhood each of us progresses to a particular stage of emotional and moral development, of which there are several stages. Then we become stuck at that level forever. We are, however, able to reason one stage beyond where we are stuck in our ability to act. If one is unaware of this layering, one cannot communicate with another person at a different level or from a different moral stage of development, because neither will understand that, in fact, they are saying and hearing totally different things.

Later in very different work, Norman Vincent Peale[4] provided self-improvement alternatives and suggested that individual optimism had great potential for impacting one's personal behavior, health and perception. To some extent, Peale proposed this optimism affected one's larger family, even to the level of the nation, continuing on up to global outcomes.

The self-awareness movement emerging in the same time frame was motivated by counseling and therapy and most certainly led us to the outpouring of self-help books throughout the English speaking world. I cannot find data suggesting the efficacy of these books, but the bookstore shelves full of them suggest one form of data that indicates value in the eyes of the average American. This type of book, where personality modification is targeted and improvement sought, is unprecedented, absent from all previous centuries. There were "how to" books much earlier, including the Bible, Koran, writings of Confucius, Plato's work and Shakespeare's plays; all of which suggested preferred and failed ways of living life. The thousands of books suggesting how to change our emotions and our behaviors are unique to this past century and will likely remain so.

20th Century Self-help book topics have included:
Abuse
Anxiety and Obsessive Compulsive Disorder
Bereavement
Confidence, Self-Esteem and Assertiveness
Depression
Eating Disorders
Families
Health (Addictions, Insomnia, and Unplanned Pregnancy)
Maximizing Your Potential
Relationships and Friendship
Self-Counseling
Sexual Relationships
Student Life
Study Skills
Women's Issues

Finally, by grouping Naom Chomsky's work and publications explaining syntax and language as a large-scale innovation of significance for improved self-awareness, we began to understand the underpinnings and limits of human communication and all they encompass. Chomsky deserves to be more present in the psychology and philosophy category than communication as defined in this book because his work is not about the "electronic communication" of the 20th Century, but in a category that gets to the root of all human discourse.

Through his work we begin to understand the importance of ghetto street languages, which develop as idiom amongst various classes within a society, and which are at times looked down upon by other classes. An example of this is the style of speech used by Blacks and Hispanics in Los Angeles, within their own groups as a means for conveying camaraderie. This language style produced, as its grandchild, Rap music, a totally altered form of musical style that comforted African-Americans in their search for dealing with the violent chaos around them.

Through Chomsky's work we also realize that the Bible, Koran,

Mormon Book of Maroni or any book from God, cannot fully and perfectly convey meaning in a language that is not perfect in its ability to convey information. We begin to understand that arrogance about one's dialect is specious. We finally realize that, without language, our ability to stay ahead of previous generations would be severely limited. Language is the vehicle for cultural memory, more important than DNA.

The understanding of language and its limits certainly led three architects to write *A Pattern Language (1977)*, a book in which they attempt to alter our way of verbalizing architecture in order to force us to change our way of thinking about the shape, form and materials used in the creation of buildings and towns. [5]

The two introductory poems in this book emphasized the possibility of common language being central to improving our global understanding of each other, in spite of its inherent imprecision.

PHILOSOPHY

In reviewing the impact of language and syntax on philosophical thought of the 20th Century, we find few but significant contributions. The most influential philosophers concentrated on mathematical rigor, including the impact of quantum mechanical physics in their discussions of reality.

Bertrand Russell asks "Can human beings *know* anything, and if so, what and how? This question is really the most essentially philosophical of all questions."[6] "More than this, Russell's various contributions were also unified by his views concerning both the centrality of scientific knowledge and the importance of an underlying scientific methodology that is common to both philosophy and science. In the case of philosophy, this methodology expressed itself through Russell's use of logical analysis. In fact, Russell often claimed that he had more confidence in his methodology than in any particular philosophical conclusion.[7]"

Bertrand Russell suggested we focus philosophical examination on the interesting problems of day-to-day human life, rather than on remote abstraction. Jointly with Alfred North Whitehead, he published a guide that connected mathematics to logic and presented proofs for theorems used in mathematics today. He then presented a series of logical/mathematical constructs that he thought should be used to define philosophical problems and answers in the language of symbolic logic. Primarily, Russell suggested we have the ability, in an age of science, to use science as a primary vehicle for understanding, and to know its limits by asking questions it cannot answer. As articulated by physicist-philosopher Karl Popper, science is central to the work of philosophy and knowledge. In spite of their respect for science, these same philosophers pointed out the weaknesses in the scientific method all of us learned in high school. It is a shame that we have no such introduction to the philosophy of science in the early years of our educational systems.

The 20th Century philosophers Wittgenstein, Sartre, Camus and Rand led us into an objectivist existence-based self-awareness rooted in science and objective reality. This century's instruction in morality stems from their use of moral relativism, a concept that evolved out of existentialism.

Finally and philosophically I add Antoine de Saint Exupéry's classic, *Le Petit Prince*. I add this, somewhat out of proportion in importance when compared to the other works, though I believe it is a central interpretation of human behavior and its variants. The story walks us through the little prince's reasoned, questioning, fact-finding hopefulness and child-like fundamentalism. His simple investigation can provide us with a place to stand during our lifetime search for meaning, truth, friendship and accomplishment.

In summary, our present philosophy is embedded in science and objective reality above all else, and contains a dash of our emerging awareness that no amount of science can answer all questions. The closer you look, the less clear some answers seem. As in the movie *Blow Up*[7], apparently a certain amount of mystery will, happily, always exist.

TECHNICAL UNCERTAINTY

Science was the sacred cow at the beginning of the 20th Century and may still be. Its irrefutability was significantly altered and its credibility softened by the events listed and somewhat amplified below:

1. The existence, for one hundred years of the famous "twenty plus unsolved problems" introduced by David Hilbert at the mathematics

conference in Paris in the year 1900, has had a profound effect on the mathematical community; it was an exhilarating call to arms. Many of the problems have been addressed and partially or fully solved. Some remain unsolved today!

2. The event of sinking the engineering marvel — the "unsinkable" Titanic, a ship with double hulls and watertight compartments, that sank on its maiden voyage as its captain sailed toward a record transatlantic crossing, created a terrible hole in technology's credibility. The Titanic hit an iceberg at an unexpected location, ripping a huge gash in the side and breaching critical sea-water isolating compartments, sinking the ship and condemning over half of its passengers to death within the span of two hours. This disaster and others — the shutdown errors at the Three Mile Island nuclear electric power plant (which led to venting of radioactive gasses to the local community), the reactor core meltdown after a reactor test induced a steam explosion at Chernobyl, Belarus, (which resulted in the evacuation of a quarter million residents) symbolically ripped the lid off science, and opened up the possibility of error in engineering. This quicksand damaged people's trust in science.

3. Finally, the existence of fundamentally unsolvable mathematical theorems introduced by Kurt Gœdel, combined with Physicist Karl Popper's elaboration of certain difficulties with corroboration of any given scientific theory, both developments of the 1930s, eroded the assumption that science was a better tool than intuition for solving problems. This vaunted science appeared to presented only part of the answer, resorting to painful experience in order to complete human understanding.

And so we find them — philosophy, technology, industry and mathematics, all interacting like chemicals in a tube.

These events joined finally to diminish the inestimable value of the new mathematics of power-series approximation that allowed computers to solve thousands of previously unsolvable problems in science and engineering. These disciplines, which had given this century its strength and life, now had to coexist with science and mathematic's own, nearly human weakness — uncertainty and falsifiablility. At the end of the century a series of books and movies depicted science spawned robots that turned on their creators, instituting several versions of genocide giving laconic balance back to a world in need of humor.[9]

NEW TOOLS

One of the century's most significant events from these later chapters was the discovery of a fundamentally new tool, not unlike the apes discovery of a bone-hammer in Kubrick's movie *2001: A Space Odyssey*. Twentieth century humans invented *the computer*. Put up against all previous tools, it is humankind's most powerful tool, exceeding even the axe/hammer, moveable type and the fossil-fuel engine. With this computing engine, with this tool which our mind can use like a crowbar to pry knowledge forward, to find our own limits as we seek ways to apply it to nature's secrets. We now use computers to reprocess CAT scans of brain images to tell us where various sensory stimuli are moving thoughts through the brain's various pathways. We guide aircraft using these computers, we detect potential collisions between aircraft, we reserve seating on those same flights, we automate factories and manufacturing processes, we evaluate weather patterns, we play chess against a grand champion, we in fact compute to excess.

This compute engine exceeds all other engines yet invented. From the engines for farming in earliest times where man was the thinking component to the horse and plow, through the industrial ages of steam and fossil fuels which lead to machinery driven farming and manufacturing, to the electric engine which finds itself ubiquitous in its use for automobile windows, refrigerators, clocks, all forms of kitchen appliance, gas station and refinery pumps, the compute engine promises more. This new engine has taken the abstraction of mathematics and moved that power into many facets of our lives. Each of the named people and devices in this short list emerged only with the support of many, many contributors. This century's growth of electronics surely made the computer's impact possible. There may have been inventors earlier who described a computer or thinking machine using the tradition of science fiction's faster than light travel, imagining a machine capable of thinking, but I place tend to place less importance on a machine decscribed on paper than on one implemented. Such ideas sit for a long time on the barren plain of disuse. Leonardo da Vinci drew airplanes and tanks in the 1400s. He is given credit for seeing the future, but had no more to do with their use in the 20th Century than science fiction writers had of rockets in the previous century. It is true that without vision the people perish,[10] but without the thousands of implementers below the surface in the development of the compute-engine, it presence would not have changed the world.

LEARNING AND TEACHING

We are now late in the book and though this section's title is broad in scope, a lot related to learning and teaching has already been covered in earlier chapters. To a large extent only what is listed for this chapter will be discussed here. We already spoke of the psychology of learning as well as the growing number of college educated members of society and the large number of higher education institutions. This small chapter will cover significant additional elements of learning and teaching not yet touched on.

New in this century was the millennia-spanning history-based fiction novel. It was probably most visible early on in James Michener's *The Source* and *Hawaii*, where the beginnings of Judeo-Christian life are unveiled first at their crossroads in the Middle-East and second through the profound effect which the Christian invasion of the Hawaiian Islands had on their inhabitants. Since publication of his twelve historical novels, many other writers have published historical novels of which Jane Auel's *Clan of the Cave Bear* is an excellent example, providing us a view of our beginnings and clarifying our place in the perspective of life over thousands of years.

The most relevant form of 20th Century prophecy also emerged from novels. In the novel *1984* by George Orwell describes a technology-enabled government of overseers who spied on, and restricted people's lives, threatening our hard-won freedoms. Those prophecies did not materialize in the year predicted, but seem to be acute warnings to us as we move into the present century. In Huxley's *Brave New World*, we fall from grace into a drug-enabled, entertainment seeking, and self-indulgent civilization. Such stories serve up warnings, waving us off paths that might lead to terrible futures. The prophecy novel became so ubiquitous in the 20th Century it encouraged the production of large numbers of books written with wild and false predictions of financial ruin and other forms of demise just waiting to disable the industrialized world. These books perhaps mirrored, or exploited, the chaos in our minds. When the outside world changes faster than we can change inside, we become frightened and confused, and may project that inner fear and confusion onto the world around us.

Doctor Seuss, on the other hand, both lightened and enlightened our vision of the future through children's stories in the late 1950s. In flowing and rhymed verse, he wrote of Star Belly Sneetches, a paunchy yellow animal that ridiculed those whose bellies did not have stars. To make such trivial separation laughable, his cat-in-the-hat sells each Sneetch a ticket through his star-on or star-off machines, raising their class status in an endless cycle of foolishness, until finally the lack of difference becomes all too clear. Stars, color, do not matter! This same wasteful racial discrimination scene has played out many times and in different places and countries, over the span of the 20th Century. Making judgments based on such easily transferable differences must end.

· · · · ·

The seven summary contributions named at the beginning of this chapter affected human beings worldwide. They forever raised our social awareness and our concern about government excess. They provided new methods of passing tribal knowledge to our children. And as I have said before, countless other authors have spoken broadly and grandly about the impact of these late chapter events and inventions.

A concept that emerged for the author during the research for this book is centrally relevant to this chapter. It is the idea that society now needs *synthesizers* to help it flourish. That is, people who take the very narrow and specialized knowledge in many fields and synthesize them into one concept, unifying many different educationally deep areas of human knowledge to make it readable and understandable. Many synthesizers already exist and our need for them is evident as shown by the popularity of hardback non-fiction and fiction science books such as the several "big bang" overviews of physics, Michael Crichton science fiction novels and the many Michenerian historical novels. In many ways the synthesizer is a person very much like the philosopher of old and very much in line with Wittgenstein's later thinking of the philosopher's assignment: "What can be said at all can be said clearly; and whereof one cannot speak in that way one must be silent."[11] The most important aspect in the emergence of new truth tellers for modern society is that they must have extreme integrity and depth.

Reference notes
1. www.biozentrum.uniwuerzburg.de/genetics/behavior/learning/ SkinnerBox.html; www.bfskinner.org/bio.asp.
   These are two web references that summarize the topic well.
2. www.wwnorton.com/psychsci/ch17_overview.htm
3. Richard J. Herrnstein, Charles Murray, *The Bell Curve* (Simon and Schuster, NY, 1994).
4. Norman Vincent Peale, *The Power of Positive Thinking* (Simon and Schuster Books, NY, 1952).
5. Alexander, Ishikawa, Silverstein, *A Pattern Language*, Oxford University Press (Oxford University Press, NY, 1977) ISBN: 0195019199.
6. Bertrand Russell, in a letter to Lady Ottoline Morrell dated 13 December 1911, quoted in John Slater, *Bertrand Russell* (Bristol: Thoemmes, 1994): 67. http://plato.stanford.edu/entries/russell/#RWL.
7. Irvine, A. D., "Bertrand Russell", *The Stanford (on line) Encyclopedia of Philosophy (Fall 2004 Edition)*, Edward N. Zalta (ed.), http://plato.stanford.edu/archives/ fall2004/entries/russell.
8. M. Antonioni, *Blow Up*, (Movie, Italian and UK-English movie).
9. Movies: *I, Robot (2004)*; *AI* (2001); Books: Isaac Asimov: *The Caves of Steel* (1954), *The Naked Sun* (1957), *The Robots of Dawn* (1983), *Robots and Empire* (1985), *I, Robot* (1950), *The Rest of the Robots* (1964), *Robot Dreams* (1986), *Robot Visions* (1990).
10. Proverbs 29:18, Holy Bible — The Old Testament.
11. L. Wittgenstein, *Tractatus Logico-Philosophicus, 1921*, in the foreword.

CHAPTER 8

Table 17   Dwellings and Structures

CHANGE EVENTS	YEAR	CHANGE AGENT
American Architecture	1909	Frank Lloyd Wright
Panama Canal	1914	United States
The Skyscrapers	1923	Charles E. Jeanneret/Le Corbusier
Functional Architecture	1924	Walter Gropius
Aswan High Dam	1960	Egypt

Table 18   Government

CHANGE EVENTS	YEAR	CHANGE AGENT
16th Amendment Personal Income Tax	1913	US Congress
The Trial	1915	Franz Kafka
Volstead Act repealed prohibition	1933	USA people

Table 19   War

CHANGE EVENTS	YEAR	CHANGE AGENT
World War I End of Colonialism	1914	France/Germany
Failed Marxism (policies)	1926	Joseph Stalin
Nationalism	1939	Wane of Imperialism
World War II triggered	1939	Adolf Hitler
War General's General	1944	George S. Patton
Atomic Bomb — Hiroshima	1945	Robert Oppenheimer
Marshall Plan Bretton Woods	1945	George Marshall
United Nations	1945	Woodrow Wilson
Truman Doctrine	1947	Harry S. Truman
Tough Minded Democracy	1951	Winston Churchill
Nuclear Submarine — 1st Strike ended	1954	Adm. Hyman Rickover
"I Have a Dream" Civil Rights	1963	Martin Luther King Jr.
Gulf of Tonkin Resolution	1964	US Congress
Peaceful prostest, UC Berkeley	1964	Mario Savio
Vietnam War — Guerilla War	1964	France, USA
Imagine — Music of the Heart	1971	John Lennon
Genocide	1975	Pol Pot

"The population centered in metropolitan areas has risen dramatically from 20 percent in the late 1800s to 70 percent in the 20th Century..."

DWELLING AND STRUCTURES
GOVERNMENT
WAR

I n this final chapter I bring together three categories not naturally joined: buildings and architecture, big government, and ultimately ... war. They are the three categories with the smallest count of unique and revolutionary changes affecting multiple millions of people during the time covered by my list. These three topics do, however, have a common thread, which arises out of the large numbers of human beings involved in each. Changes in building design and construction initiated by new materials and technologies resulted in the growth of huge cities, concentrating most of the 20th Century's population in metropolitan areas; government, the single largest employer in most industrial countries and thereby the provider of a similar human concentration; and the wars, involving the world and killing over seventy million people — certainly a concentration of human suffering. When we view the century through those three windows of centralized control of millions of people, we will notice that 20th Century wars were not the most significant element of our century in terms of changing the future of human life. Others may argue that they preserved that future, but that is not the focus of this story. War is no longer the only agent of powerful change.

The Dwellings and Structures, Government, and War based events, discoveries, inventions and changes of the 20th Century certainly made their impact on society. Their most lasting effects were:

1. Major changes occurred in the design and materials used for dwellings and public structures. These changes most markedly affected high-rise office and apartment buildings, huge dams, and public cultural and sports arenas. Skyscrapers were a truly thousand year phenomenon of the 20th Century, creating city densities beyond those of all former centuries; a Los Angeles of 10 million people, an agglomeration around Tokyo of 34 million.

2. Unprecedented increases in the size of the U.S. government and of industrial nation governments worldwide. This ubiquitous presence of government as employer, protector, warrior and tax collector became a unique new element in the world.

3. Democratic moralism, using majority public legislation to limit indulgence, failed the effectiveness test, though that form of moralism has not yet been abandoned. Hopeful restrictors continue lobbying. (See Chapter 1).

4. War became a smaller player in world events of significance compared to earlier centuries, especially when viewed from the point of view of land boundary and leadership changes. Most of those changes were made through democratic processes. In spite of World War I's and World War II's human toll, and the horror of the killing of innocent Jews, few boundaries were permanently altered.

5. The boundaries that were changed in the 20th Century played a major role in ongoing conflicts such as those between Israel and Palestine, those along the Slavic peninsula, in Southeast Asia and in India and Pakistan.

6. Contrary to some popular beliefs, technology did not gain most of its ground from the coffers or pressures of war.

7. Ghandi invented non-violent passive resistance as a tool for creating non-violent social and governmental change. That tool was used to

liberate India from the British, African Americans from segregation and students from dictatorial university rulings.

Dwellings / Structures

Though this list is not extensive, it illustrates key structures and architects that had a major impact on multiple millions of peoples' lives, such as the dam that allowed control of Nile plain flooding, the canal that allowed cargo transfer from the Pacific to the Atlantic without long and hazardous Cape journeys, and the many skyscrapers and sports domes that populate America's modern landscape.

This was not the age of classical monolithic structures in general,

Panama Canal zone (note person center foreground)

nor was it the era of building the most important elements in classical cities. It was the age in which steel-reinforced concrete, patented just before the 1867 Paris World's Fair, combined with the improved steel making processes of Henry Bessemer in 1855, and also combined with the use of the load bearing structures of modern high-rises developed by George Fuller in 1889, revitalized the construction of taller and taller buildings with wider spans than ever before accomplished in all of human history. The tower of Babel pales to insignificance (diminishing in no way its symbolic power) compared to modern high-rises. An interesting present-day comparison exists between low-rise Paris, still a classically built city of great nobility, and the overarching height and density of the buildings on both the island of Hong Kong and the island of Manhattan. It's interesting to me that steel-reinforced concrete, a relatively recent combination of two common and ancient materials, is so effective and useful.

The Rise of Cities — Hong Kong 2005

The primary value provided by the high-rise office building was dense white collar office space in commercial cities, those near a port for trading goods, and side-by-side with those offices were apartments for living in equal density. Thus, to the extent that adjacency of working groups from different and like businesses creates new efficiencies and new ideas in human commerce, these cities have the potential for changing our lives in coming centuries. The populations centered in metropolitan areas have risen dramatically from 20 percent in the late 1800s to 70 percent of the country-wide total for industrially developed countries in the 20th Century.[1]

Private homes were reshaped by many architects of the 20th Century. Frank Lloyd Wright most famously illustrates that facet of American home elaboration. The arts and crafts movement in England and later in the Western United States combined some of the elements of Wright with the Japanese style and integrated the warmth of wood and light to create the more opulent western homes of the 1930s. The first tract homes (17,000 of them), were created by William J. Levitt[2] in Levittown from 1952-58. He set a new standard for the rate of growth for new suburbs (moderate density cities adjacent to large cities) popular in the post-war boom of the 1950s.

GOVERNMENT

The 20th Century has seen significantly larger growth in the size of government, promoted as much by the invocation early in the century of the personal income tax as by industry and population growth. Taxes in various industrially developed countries vary from slightly under 30% in the U.S. to over 50% in Sweden. Government then expanded to fill the size of its budget. That growth was augmented by a growing nationalism centered on the two world wars. By the end of the 20th Century 34% of the GDP was due to federal and state government expenditure, 22% of all jobs could be said to involve federal, state and local government employees. The GDP over that same time span was 63 times larger than it had been in 1899 (and even larger if calculated in equivalent dollars called Chained Dollars on the U.S. government's Bureau of Economic Analysis web sites).[4]

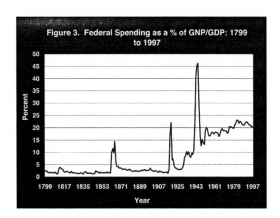

Figure 3. Federal Spending as a % of GNP/GDP: 1799 to 1997

In response to this ubiquitous governmentalism in Germany, Franz Kafka wrote of the insensitivity of the government machine in his book, *The Trial*. That insensitivity needed no more evidence than that evinced by the House Committee on Un-American Activities in late forties

America. The HUAC's publically accusatory questioning of suspected Communists engendered publicity sufficient to damage the lives and careers of the guilty and innocent alike, without benefit of trial.

In 1919, the Volstead Act, the 18th Amendment to the United States Constitution, prohibited the sale of liquor across state lines. The year 1933 saw the repeal of that same act and with it, a retreat from placing responsibility to regulate human indulgence through the government's power to make activities illegal (This topic was discussed in Chapter 1).

In opposition to big government, Congress enacted the Capper-Volstead Act, giving farmers the right to join together in cooperatives that grow, market and sell their own products. This flew in the face of anti-trust protection of the consumer by protecting the small farmer, because they were considered essential to American economic health.

The presence of government is, at the same time, a vital necessity to the rule of law, to the provision of industrial infrastructure and for protection of citizens from themselves and other nations. It also appears that the size of the coherent nation ruled under a single democratic leader and without difficult internal trade restriction is vital to worldwide competition as evidenced by the European Union's recent increased bargaining power.

The 20th Century has had to embrace a worldwide authority in the body of the United Nations, but that body is as yet insufficient, though it is an independent peacekeeping body as well as an adjudicator of international dispute. How the world moves toward a nationalism that is internationalism will be our assignment for coming centuries.

WAR, DEATH, AND HUMAN CONFLICT

In earlier centuries wars and their subsequent boundary changes seemed, based on reading history timelines, to be the primary change agent. To appreciate the importance of war in history, the reader need only take a look at any of the books listed in the selected bibliography.

In stark contrast, wars of the 20th Century, though filled with horror and death, did not dominate the list of causes for the changes wrought on Earth's millions by the century. Multiple millions were killed and injured in the wars of the 20th Century, and my intent is not to diminish the sacrifice of those men and women, nor the historical impact that would have resulted had the outcome of either of the world wars been different.

And certainly World War I's end placed a last nail in the coffin of Europe's centuries of colonization.

Others want to suggest that World War II was responsible for everything technological in the 20th Century, which cannot be the case. World War II did foster the invention of radar, but the tubes for radios and for audio amplifiers were invented before World War I; by 1916 in fact; and loudspeakers by 1924. They say science is its own reward. Curiosity, the natural human desire to *know*, renders even war small as a driving force toward discovery. At least that is the legacy of the 20th Century. The material science for transistors was invented at Bell Labs, a public utility company regulated by the government at the time, funded by public telephones, not by the government or the war. Without transistors, the communications chapter of this book would not have been written.

The world wars did bring technological change to weapons (atom bombs), to radar, to submarines (unlimited submerged operation), to aircraft design and to jet and rocket engine development as well as enhancement of warship designs. Commercial enterprise remunerated jet flight surely, though the other areas of war driven development continued to be government and war funded.

The wars were a part of history that saw millions die at the hands of despots and their sad visions. There is so much written about 20th Century wars — entire bookcase sections at bookstores — that I will not add to this vast collection, nor attempt to review its details. I refrain from suggesting what those wars meant or will mean to future wars in future centuries.

On this topic I have chosen to limit myself to summarizing the most relevant lessons of this past century's wars.

1. That failing to provide support for economic recovery of war torn nations results in an unnecessary repeat (WWII) of the same war.
2. That weapons will keep getting more destructive, until they endanger not only their intended targets, but also their inventors. Human leadership will have to manage and control them in order to prevent

our own extinction. We now know the awkwardness of having the atomic bomb, a weapon with the power to destroy whole cities and potentially, with the help of nuclear winter, even nations. We also know that nuclear electric power uses the same technology to provide our energy needs, a critical "food" for our coming century.

3. That winners drawing new boundaries do not end boundary conflict, a truth well illustrated in the creation of Yugoslavia, Czechoslovakia and Israel at the end of World War II.

4. That creating a world leadership council, such as the UN, is not an easy and always effective process, and is made harder by the disparity in levels of industrialization and poverty. And that it is a critical ingredient to world cooperation.

5. That inventors like Martin Luther King Jr. and Gandhi can turn masses from using violence for motivating change to passive resistance that foments change with less loss of life.

6. That individuals such as Hitler and Pol Pot can kill millions in genocidal acts of violence, similar to nature's 1918 flu, a virus that also killed millions mindlessly (flu deaths of more than 18 million).

7. That the re-emergence of guerilla warfare (invented in the American Revolution) and its natural evolution leading to suicide bombers who walk into large human gatherings and detonate themselves to create terror, is an emerging means of warfare yet to be understood and countered effectively and whose one sure consequence is to reduce individual freedoms through over-governance in the coming

Honoring The Fallen - Arlington National Cemetery

century.

8. That the reawakening of the world, and in particular the United States of America, to the impossibility of forcing a form of government on people willing to resist, as in Vietnam where modern guerilla warfare, now lately coined insurgency, came of age.

9. That the alcoholism and drug addiction common among the soldiers who kill in hand to hand combat stamps its sadness and guilt onto the families and future families of those soldiers through several generations, extending the abuse and horror of war.

10. That war seems to be an inevitable activity in each century and understanding its use is important, even in advanced societies.

If you look at any historical timeline from the beginning of civilized groupings of people, and their assemblage into city-states with armies, what stands out on that time line is the significance of war and its role in boundary and leadership changes. Both war and sickness can halt civilizations' advances, as occurred in the time of the Black Death and during the Dark Ages in Europe. Among all the events of the last four thousand years, it is this fact of war being central to the history of that time which has most marked time prior to the 20th Century.

War, however, gets second place in the confluence of significant events of this past century. War still gets the award for most people killed under a common umbrella, though disease, starvation and environmental calamity deserve a strong second place. It appears that for once war has taken a minor role in human progress (or regression). And in that progress we have learned many lessons. We have been admonished and to a degree we have heeded the warnings. We have also gained at least two significant examples of peaceful revolution in the persons of King and Gandhi, and many examples of peaceful change in leadership using the universal tool of suffrage.

This reduction in war's relevance has occurred despite the ungainly size of government in modern industrialized countries. Does this not suggest a basis for hope? We must continue the emphasis on non-violent settlement of conflict, but must also continue to be prepared for war when despotic would-be-kings arise, as did Hitler in 1938.

Happily we have not yet encountered the horrors of large bureaucracies envisioned by Kafka — that government will become an

144 NOT IN A THOUSAND YEARS

unhearing behemoth meting out injustice more frequently than justice. As the need for worldwide governance increases, there will be a need to restrain government growth if we are to continue in an un-Kafkaesque way. We must not allow our voting rights to atrophy.

And finally, within our massive cities we must find ways to stay connected to one another in human warmth and friendship. If we can do this, then from these monoliths of work and thought will arise the coming world's answers.

Reference notes
1. National Geographic Society World Atlas, 2004.
2. Pennsylvania historical web site: http://server1.fandm.edu /levittown/one/b.html.
3. 1998, www.house.gov/jec/growth/govtsize/govtsize.htm.
4. www.bea.doc.gov/bea/dn/home/gdp.htm.

CHAPTER 9

"The need for synthesizers as the new philosophers is a social need of paramount importance if we are to see the forest for the unbelievable complexity of the trees."

SUMMARY AND CONCLUSIONS

I began by promising to reveal our century as one of unique and lasting changes similar to those to art during the Renaissance and books after Gutenberg's moveable type. Together we walked through a narrative of the events that affected multiple millions of people. Although our list was large, it constituted only the tip of the iceberg of events. We fleshed out the sub-categories of this collection to make it mentally more manageable. We then collected the items in each sub-category into their generalized impact on society and our world, elaborating on each of them a little to improve our understanding.

Now we can summarize those changes that had the most lasting impact and to decide for ourselves, in that boil-down, whether our century deserves to be placed alongside former greats like the renaissance and the democracy of Greece, or relegated to the also-rans.

I have broken this summary into four groupings:

- The changes that increased our *reach*.
- Those *technological* changes affecting our operating lives.
- Those changes altering our *cultural-moral* framework.
- And appropriately last, our awareness of *limits*.

Reach

The societal flow of technological progress in the 20th Century pivotally changed from the previous era in which one idea changed the world. The 20th Century is very differently characterized by thousands whose ideas changed the world. My list is only a symbolic sampling of those thousands who participated in the changes accomplished through human achievement during the 20th Century.

A major new tool emerged — not just a tool reinvented or made more automatic, but a *New Tool*, the computer and its compute-engine, the microprocessor. This is a tool that extends human understanding and enhances the intelligent automation of tasks. Its invention was at least as important to humankind as such earlier new tools as the piston and electric motors and the printing press with moveable type. The compute engine's ultimate influence is still unfolding and its value may become as large as language itself. To open one's laptop, no thicker than a half ream of paper, is to enter the library of the world.

For the first time we have become an *Earth-wide society* with electronics-filled satellites that provide weather, telephone, television and global positioning information that can be obtained anywhere on the planet. There is no place on Earth that cannot choose to overcome its isolation. Low-cost, worldwide communication and access to information has collapsed distance and in many ways made obsolete some uses of transportation so recently implemented during our century.

That same *planet-wide transport*, however, provided access to countries, goods, information and people. The worldwide availability of images, movies and music enabled a new *universal language*, ending the era symbolized by the Bible story of the halted construction of the Tower of Babel. Jazz, rock and roll and movies became the vehicle for worldwide understanding and communication beyond language, a vehicle that moved us toward global unity. A mathematical understanding of language was established that will eventually lead to compact, machine operated translators (some forms of which already exist on computers). These same global messengers, movies and music, could also shock and enrage certain cultures and are one of the causes of terrorism.

Planet-wide transport also provided goods from low labor-cost nations to high labor-cost nations. This worldwide labor pool's availability induced local upheavals and economic disruption as well as improvements to living standards along with the unintended cultural influences on the selling nation.

Unprecedented increases in the *size of governments* worldwide altered governmental structures and strictures. The government became a larger protector and warrior, though the early century dominance of the military budget dropped significantly in the later years of the century, giving way to subsidization of the aging population's health and to income support of the disabled. This ubiquitous presence of government as both employer and taxman became a unique new constant of the "free world," consuming upwards of 35 percent of our earnings. Fast, closed-loop, societal feedback systems enabled quicker government and business corrections to ill-conceived plans, allowing those plans to be made and altered quickly, sometimes too quickly. With government as our warrior, government and science united to create the *awesome destructive capacity* of nuclear weapons with the possibility that their use might severely alter life on Earth, if we are unable to control them.

As part of our inward-reach we increased our *understanding of the human body* a hundredfold through advances made using the tools of science. Over 95 percent of our understanding of bacteria, viruses and infection occurred in the 20th Century. We began to prevent and cure diseases by vaccination and antibiotics. Molecular biology's chemicals provided sophisticated abatements for previously incurable diseases such as AIDS. The discovery of DNA brought us to the realization that we now understand the fundamentals of human life. Engineering's creation of imaging machines now provides far more non-invasive body-diagnostic techniques than ever before. We have mastered complex surgical interventions and created electro-mechanical implants that repair body parts when they break or fail from age. All of this has led to a huge economic burden on society for health care, now exceeding 12 percent of GDP in the USA. This fascination with health led to an increase in size and the social influence of an older population that is just beginning to make its mark. It is possible that the power of, if not the respect for, the wisdom of the elders will be reestablished following decades of decline in the West.

I have stated categorically that *literature* reached its peak of achievement in the 20th Century, its golden age. Offset printing processes provided the most perfect and economic books of all time and television had not yet shortened and usurped the attention span of modern readers. The historical novel emerged in part as a means of extending our ability to remember history in order not to repeat it.

Abstract fine art was invented in the 20th Century and, by borrowing new materials and inventing new forms and venues for its expression, left a significant and popularly unacknowledged impact on the future of representational art as well as on all artistic legacy.

Finally, *engineering design and manufacturing process* technology matured and may have peaked. What has not yet matured is society's ability to make the proper choices for design requirements. *Society must decide* on durability vs. cost and style. Do we want a lifetime automobile that costs more than a house or one that is renewed with technology and design tastes every three years, but breaks if you keep it too long?

The maturing of the design and manufacturing processes that provide highly complex products, optimized to meet consumer and society needs, is a recent and astounding phenomenon. This design maturity allows costs, durability, function and beauty to be managed during the design phase, providing people access to products largely independent of cost. Using modern manufacturing and design processes it became possible for a small number of people to produce millions of units of any device. The magnitude of this change is similar to the advantage provided in earlier centuries by the invention and use of the combine for harvesting grain for food or the cotton gin for clothing, and possibly, much larger.

Technological Effects

Countries began to use *Keynesian deficit-economics* to support nationwide economic recoveries. This was unlike the credit-conservative lives of the individual. This adaptation has appeared to work well in avoiding a replay of the worldwide economic depression of the 1930's.

The *knowledge-worker* emerged as a result of a major increase in the number of universities during the 18th and 19th Centuries and began to drive the structure and accomplishment of working society. Countries employing large percentages of knowledge workers began to outperform other countries. That led to an increased demand for education systems that provide such technically trained employees.

The emergence and demise of Communism as an economic and political system took place. The void was filled by worldwide economic *democra-preneurship*, complemented by an increase in government-managed support of the unemployed and the aged, with its resulting social costs. The social costs experiment has not yet stabilized and will have to be dealt with in the present century.

Energy became a prime ingredient to industrial society's survival. The creation of the electric utility, combined with the broad-based availability of household appliances and the mass use of the automobile, drove our need and desire for the energy products of gasoline and electricity. These sources of machine energy became as vital to families as food.

Skyscrapers became a truly thousand year phenomenon, creating city densities beyond those of all former centuries. As revealed in Charles Murray's *Human Accomplishment,* the creative outpourings of humanity emanate from these *ultra-dense population centers.*

Movies allowed us to *experience many different lives* within our one life. Movies also projected us into non-existent worlds of the future and into no longer existing worlds of the past. Our visual experience became enriched beyond the most fantastic of our nighttime dreams.

A society with perspective and vision is new to civilization in its breadth and depth and will further accelerate the *effect of thousands*, rather than The One creator of change. The century's thousands of scientists unlocked the innermost secrets of matter so that we now understand the basic building blocks and mathematics of all matter, even living matter. Only gravity, of the many measurable forces, remains to be more completely understood.

Moral Culture

War became a smaller player in world events of significance compared to earlier centuries, especially when viewed from the point of view of land-boundary and leadership changes. Certainly the world wars spent money and lives as never before, but the long-term effect of those successful wars will not attain the significance or influence attributed to wars in earlier centuries. Those few boundaries that were changed played a major role in creating ongoing conflicts.

Gandhi invented non-violent passive resistance as a tool for creating *non-violent social and government change*. That tool was used (for the most part) to peacefully liberate India from the British, African-Americans from segregation and students from dictatorial university rulings.

Women and African-Americans, for the first time, claimed the *right to vote and work as equals* with white men. This doubled the workforce and changed the paradigm of raising children while both parents worked. Developing countries also came under pressure to institute significant changes that support fairer labor and democratic practices.

Democratic moralism on the other hand failed. Its use of majority public legislation to limit individual indulgence via prohibition collapsed two short years after its inauguration in the United States. Sadly, that form of *moral imperialism* has not yet been abandoned. Hopeful restrictors of indulgence continue their societal lobbying. The experiences of the century suggest that the best moral compromise is often reached by working on the problem and not creating bimodal self-righteous distributions of people.

None-the-less, *structural social changes* have begun — gender-racial, moral-sexual and government-economic. A relaxation of sexual mores accompanied the use of pregnancy prevention outside of marriage. An increase in the ease and frequency of divorce also altered marital-sexual consequences. These developments presage further changes in our socio-sexual culture. The changes will be vast, but will take time, probably a century or more. The cultural consequences to our society once we are individually changed will take even longer to assess. The slowing effect I described in a social culture's morality response-time results from our inability to forget our generational pasts conveyed to us by language and parental behavior. That past appears to require about four to ten generations for lasting effects of social behavior modification to stabilize — up to three hundred years for implications and corrections to be concluded. Our present day integration of races and genders with an overwhelming array of knowledge and perspective cradles us for more informed participation in our future. A worldwide sense of fairness and possibility greets the twenty-first century.

Mathematics and science dominated the direction and evolution of philosophical thought. To some, this further withered the philosophy of morality so popular with Thomas Aquinas, who, in the 13th Century lived "at a critical juncture of Western culture when the arrival of the Aristotelian *corpus* from a Latin translation reopened the question of the relation between faith and reason . . ."[1] The need for *synthesizers* as the new philosophers is, as suggested in the text, a social need of paramount importance if we are to be able to see the forest for the unbelievable complexity of the trees.

And finally, not visible on the list of contributions, are those eruptive forces not caused *by* contribution or change, but *in reaction* to them. The ubiquitous presence of music, movies and the Internet caused a reactive force in cultures that resist the influence of these new social behaviors. Such an effect is most visible in the rigid canonization of various religious formalisms such as those presently unfolding in the Middle-East Muslim community. They react to the harsher edges of western morality, such as its web-wide displays of erotic images, and provide reactionaries with a surface upon which to morally press against those changes with outrage and violence. The concept that "every action causes an equal and opposite reaction," though not a physical fact in this case, appears to have more weight than expected and it requires more careful thought when reviewing change in its now wider world context. If some control and preparation for change are not better managed, the violence of these reactions may come to dominate, and in some cases obliterate, human progress.

Limits

And last but certainly not least is our growing awareness of natural limits to human dominance and control.

The *curative therapies* for emotional and mental pathologies were primarily developed in the 20th Century and expanded by the introduction of chemical treatment. Reductions in pathological behaviors are not well quantified in the literature, so objective evidence of improvement throughout society is scarce. The pervasive presence of self-help books for all manner of human disturbance and activity, published for the first time in quantity, is a sign of significant social interest.

I noted earlier in this summary the value of the expansion of our awareness induced by movies and TV. There was a flip side in that those same media created severe *social isolation* in the neighborhoods of the developed countries. The fallout from that isolation is managed to some degree by the new psychological treatments, but more of the consequences of isolation will emerge in the present century.

Fear of catastrophe and exploitation of those fears by governments and the media to capture our attention and obedience seemed to permeate life in the latter part of the century. The sense of wellbeing in countries with high GDP has been unfairly marred by the media's exploitation of our fears regarding the potential catastrophic impact of human society on our Earth. Not that there are no potential catastrophes, but that the presenters of these stories do not substantiate their claims. They use fear to attract readers and viewers. Catastrophe is prominent in movies, in books, and in the literature of the environmental protection movement. Even our own information-conveying news organizations prey on us without discretion or evidence. We, the recipients of the media, must vote with our attention in order to *reestablish the importance of the science* behind each worst-case scenario, and to ask that they present the best-case scenarios as well, if we are to emotionally and rationally survive this new climate of fear. Constant reversals in the scientific conclusions of predicted outcomes frustrate us as a people who wish to stay informed, and these reversals devalue the credibility of our own news and information sources.

We also began to understand the *limits of science* and mathematics as tools. The science behind these limits emerged from many studies and papers. Science as god was replaced by science as tool. Human beings now see their planet in perspective with the wider universe. We are aware of the planet's limited resources and its limited ability to cleanse its own air and water. We have a dawning awareness of time that has passed since our human presence on Earth began, how races evolved in different places and what early man and woman were like. The 20th Century has provided individuals, for the first time, with a vision of our short place on the long line of time. Just as early oceanic explorers guided their ships onto the ocean, so we begin our journey into the seas of our solar system and its universe.

· · · · · ·

So much has changed that we strain the limits of our ability to adapt. Technological changes seem to provide the easiest acclimatization; cultural changes the slowest and most socially painful. We are proud of both our accomplishment and potential and much more aware of our limits as we move through this new century.

Reference notes
1. http://plato.stanford.edu/entries/aquinas.

EPILOGUE

"Biology is . . . on a threshold from which beginning of the century predictions can only suggest directions and not destinations."

IMPLICATIONS
REGARDING
THE 21ST CENTURY

I have used some of the insights gained in this review of the 20th Century as a lever to pry open the future. I place these forcasts here as an epilogue for your enjoyment, since prediction is fraught with inaccuracy, and this book's intent is non-fiction.

Prediction can provide value in two ways.

The first is that prediction capable of suggesting a general direction of technology and society which is useful in proposing aspects to study and prepare for.

And second, prediction can provide timing estimates that are accurate enough to suggest the most effective ways to use our short time on this earth, i.e. one's personal resources and commitment through investment, action, legislation, or education. The necessary timing accuracy is what makes prediction so error prone. The accuracy of most predictions are retroactive; a person is interpreted as an accurate visionary only as the future unfolds in a seemingly accurate synchrony to one's predictions. All those who mis-predicted are scorned or more likely, forgotten.

In spite of the difficulty of prediction, I wish to explain a few ideas that focus on social, scientific and cultural trends that are more likely to be present in the 21st Century's mainstream than the many other possibilities. All this, in the hope of exposing the most probable veins, as it were, to be mined this century.

The technological category most similar to the communication electronics of the 20th Century will be the consequences of our exploding knowledge of biology, its chemistry and its engineering. Let me start by examining and predicting within that specialty. Looking at the list for communication in Chapter 2, it becomes clear that the early items on the list (such as the radio triode and the telephone) would not have made us able to predict the emergence, late in the century, of computers or satellites. Biology is similarly on a threshold from which beginning of century predictions can only suggest directions and not destinations.

Genetically Modified Plants and Animals

Biological control of humanly redesigned organisms is already in the petri dish. The larger scale consequences of these experiments are pending. Investigations examining unexpected side effects such as premature aging in cloned animals are ongoing (though more recent cloning has avoided this particular consequence). Those who act as if biological and genetic modification of animals, including the human animal, should be halted because they have concerns regarding its morality are as regressive and foolish as the 15th Century pope who requested recantation of a sun centered planetary system from Galileo.

When we entered the 20th Century, we unveiled what I called the secrets of matter inside the atom. Using those secrets we unlocked the electronic and atomic age. In the 21st Century the secrets of biological matter have already been unveiled — the building blocks of DNA: its four bases, U, C, A and G.[1] Understanding these bases will unlock the biological age in ways similar to the last century's electronic age.

Genetically modified (GM) foods are still somewhat experimental and a protocol is being developed to fully evaluate the complex consequences of their genetic alteration on yield (bushels pre acre) and cost per acre, and consequences to the environment in which they are planted. Experiments are underway in which grains and beans so modified are being consumed by millions of people with no reports of negative consequences to the consumer. Other experiments have focused on building in natural pest resistance by borrowing genes from other plants with natural resistance to reduce crop loss and the spraying of chemical herbicides. China and Australia are intensely investigating genetically engineered cotton and yield improvements in grains. The best news is that genetic engineering provides similar but faster and more predictable changes to plant varieties than the old fashioned method of hybrid crossbreeding. If we can come up with a proper methodology for assessment of impact to our planet caused by such changes, we can be successful at improving our food supply.

The grand experiment of cloning everything has also begun. The vision of cloning portrayed in the movie *Jurassic Park* is presently inaccurate and overly dramatic, but it is based on science and provides a caricature of what might happen. As individuals, we can make a difference in wisely choosing a means to proceed. Rather than being a society of recalcitrants choosing to be dragged into this new reality kicking and screaming, we might formulate aggressive assessments of the best way to use our new understanding of biological modification, selecting a trusted group of science, arts, and faith elders to rationally and spiritually guide our experimentation and our learnings toward this new reality. Or we can repeat our mistakes with nuclear power wherein we became filled with fear and retreated from the cleanest means of electric power generation ever invented, until eventually we are forced to proceed with incautious rapidity due to unfolding events and paucity of oil. Protests can shut down investigation and prevent learned and cautious progress. More rational countries than ours may then surpass those reluctant to lead. I do not counsel abandoning good judgment, nor shallow evaluation of societal impact; I do counsel moving forward.

We must begin genetic modification experiments thoroughly and slowly. There are subtle side effects that have already come to light. Some genetically modified plants are not as attractive to insects as their unmodified counterpart, which alters the environment where the plants grow, making them susceptible to new prey that were previously eaten by the predator insects attracted to the unmodified plants.

Genetic engineering may also provide significant benefits to us such as vaccines that cure cancer. Early on in the development of jet aircraft some people thought transporting human beings faster than sound would irrevocably damage our bodies; an imagined consequence of early flight that had no validity, but which might have held humanity back from faster than sound and intercontinental flight for years.

Human biotechnology

On the bioelectronic front, neurostimulation using electronics has begun to show favorable results. We have seen the emergence of implanted electronics that cure body malfunctions. Electronic-neural hearing implants in the cochlea have given the deaf back their hearing (bypassing the eardrum, its bones, hair cells and calcium ions). Electronic pulses are reducing backpain using paresthesia (neurally stimulated anesthesia). Pudental nerve stimulation (at the interior base of the groin) provides electronic control of urinary incontinence. Deep brain electrodes may provide relief from Parkinson's tremor. Most recently electronic eyes for the blind bypass the lens and retina and go directly to the optic nerve. These electronic prostheses already exist. Some are widespread technologies, some emergent.

Mechanical prosthetics are also capable of curing the body's structural problems through the use of artificial arms, knees and hips. These engineered replacements make disability less burdensome for disabled individuals and for society. The early instantiations of engineered body parts were somewhat crude but effective in a number of cases. Later versions will provide functionality and durability closer to the body's original equipment.

Pharm-tech, as it is affectionately called, is a technology for formulating pharmacological constituents, often taken orally, that cure, improve or comfort us. Pharm-tech's span of improvements is large and varied, including managing health issues of the circulatory system, the emotions, and pain reduction, as well as curing or controlling acute and chronic heart disease and diabetes.

Health Care

The greatest challenge for the 21st Century is to find a means to balance the costs of drug development, manufacture and distribution with the reality of human suffering. Society cannot cure all suffering at once; our focus must be directed wherever our science and needs lead us to be most effective, given the social costs and ability of the beneficiaries to pay or to be subsidized. At the same time we must provide a for-pay service that rewards the hardworking, thrifty and rich, as well as those accidentally born into nations not as gifted with resources or industrial development as our own.

My personal suggestions for making health care affordable are:

• Continue to provide universal healthcare for the aged, but create some form of cutoff when prolonging life at high cost is no longer reasonable — create either an enforced health directive or a rewrite of the Hippocratic oath. It would be foolish to bankrupt a healthcare system because you are unable to make decisions about when to cease providing the health care of those close to death.

• Never provide national healthcare on a broad basis. If necessary, provide carefully delimited universal healthcare to children and the disabled, and require all work-capable people to fund their own health care options. Insurance makes us oblivious to the actual costs of care. This lack of awareness leads to lack of knowledge, misguided voting, and bankruptcy and needs to be fixed.

• Develop a moral philosophy for aging and educate the public to that philosophy. Politicians must become thought and legislation leaders on this subject because they should be in a position to understand the financial limits of our capitalistic democracy.

To illustrate how far the changes in the depth of biological science will go in this one century, I suggest this scene from the future: At the end of this century, catalog purchases of recombinant DNA will be available to repair known family gene defects prior to conception for those who can afford such unnatural selection. Society may find itself with an accelerated bioscience, Darwinism. Society may then have to allocate funds for DNA equality in preference to healthcare equality, simply to reduce the cost burden of weakly DNA'd individuals.

Life Utilities — Manufacturing Processes

Nano-tubes and other micro-machined and molecular constructs will provide advances in material strength per gram on the one hand and a means of chemical processing on the other (we already analyze DNA using silicon microchip-based machines). These will enable technologies for life utility devices that provide more strength per unit weight than our present supermetals (such as Titanium). In some areas these are likely to provide breakthroughs for society, in others, mere evolutionary improvements similar to those provided by Kevlar to bullet-proof vests and graphite to bicycles. Carbon-fiber (graphite) is a great material for bikes because it is strong, light, and can be made rigid where rigidity is required and flexible where that is desired. Some possible uses will be lighter automobiles, the space elevator suggested by Asimov may also assist our cost per launch into space, and simpler items such as the currently most successful MEMS (micro electro-mechanical systems) unit which powers all digital home video projectors, the DLP.

Specific function robots such as the vacuum cleaner robot and the manufacturing factory robot doing assembly and welding tasks on items from automobiles to hard disk drives will become increasingly available, but no human-like robots are likely in this new century.

Computers will continue to exist primarily as a new and revolutionary tool that enhances and deepens thought as well as simplifies or eliminates routine human work.

The greatest change for business will occur in the labor sector called *in-person services*. That sector is not outsourceable, but will continue to be a source of low wage jobs. As long as immigration provides labor for this market, low cost labor will continue to flow into developed countries just as we now know it, with human motel managers and fast-food servers. Otherwise, where it makes sense due to upward pressures on wages, *in-person service machines* are likely to replace humans. At some point, human service will become a luxury, such as it has in most telephone product support transactions where, at first, a machine directs your navigation through the choices and only after unsatisfactory answers are found, and after a wait of some minutes, do you gain access to a person. A recent example of conversion of in-person service to service-machines is the new airport check-in and boarding pass robots that replace or augment the personal check-in agent.

War, urban guerillas and terrorists

In all previous centuries the power of war and human violence, whose end is to settle irreconcilable differences between and within nations, has held humans hostage. These upheavals are similar in effect to those caused by the worst in disease, weather and earthquakes.

In this coming century a war waged by subgroups is gaining momentum via the suicide bomber. The availability of powerful technological weapons increases the lethal effectiveness of such single person terrorism. How society and civilization react and formulate attempts to control such terrorism will determine whether we reduce or strengthen these social terrorists. Technology will provide some answers to the need to track individuals, addressing the primary trustworthiness of individuals in a crowd. Redress of grievances may also reduce the level of atrocity allowed by the larger passive or terrorist supporting populace. What is most clear is that individual freedom of mobility will be one victim of such an ongoing angry-individual war. Presently religious fascism and poverty seem to be the major sources of this social malady.

The emergence of major new economic powers in India and China, long-sleeping population and geographic giants, will also play into the complexity of territorial and economic boundary struggles. World tribunals will have to grow in respect and power if some level of world tranquility at the larger multi-country level is to be maintained.

China will begin an emergence similar to that of the United States in the 1800s. China's rise could possibly mimic our early 20th Century strength if it remains as economically progressive as it appears to be at the end of the 20th Century. India will be so consumed with internal development and provision for its people that a continuation of its service profile on the worldwide market is likely. India's labor market will, however, become one of the three dominant labor pools of world order and will be the most well educated in that pool for most of this century, the other two low cost labor pools being China and Malaysia.

World wide job partitioning, the percent of people in different professions, will broadly mimic the changes that occurred in North America and Europe over the span of the 19th and 20th Centuries, moving away from the dominance of farming and toward the service industry, the professions and the sales and knowledge worker. (These changes will not likely occur on continental Africa during the 21st Century.)

Managing the consequences of the spread of nuclear weapons may become a century-dominating obsession if rogue users instigate large losses of life in advanced nations.

Web Ubiquity

The web may have gone into temporary hibernation. A major contribution from this crucible of need will emerge early in the century, though it is at present invisible. Will worldwide firearm tracking eliminate gun problems? Will web democracy change dictatorship? Will web advertising speed up the distribution of innovations in biotechnology? These are conclusions you may reach by your own review of the major contributions to the 20th Century and its implications for the 21st.

The web will become our reference dictionary, our yellow pages and white pages, our source for information. All other sources, visual and tangible, will become focused on the web or will die. The paper encyclopedia and its door-to-door salesman are gone. The web will increasingly dominate all forms of information exchange.

The easy access and display of information on the web will reduce commuter traffic peaks, provide travel information, allow review of both cost and perceived value of hotel rooms and assign commercial travel seating. Idle-time will become the focus of web-enabled reductions, attempting to further reduce standing in line by replacing it with the feeling of action. The non-web flight ticketing service computers are one example of this action oriented fast commerce, as well as the printing of boarding passes at home twenty-four hours ahead of flight time.

Hardware will continue to provide more and more portable devices for access to the web, to personal data, to communication and to the arts of music and movies. At some point a more unified set of specialty devices will emerge, but not before decades of experimentation with specialty devices such as the laptop computer, Ipod, cell phone, CD player, PDA and DVD player. Extinction models to be noticed from the 20th Century for parallels are the early dot and dash telegraph system for messages, the human plugged switchboard, the radio broadcast evening mystery program and the family card game.

It is more and more possible that online diagnosis of sickness will find its way onto the web. The office call may be replaced by the web call. Sales calls may become video-conferences where travel time is an efficiency issue.

On the darker side there are no technological or other solutions in sight for the aching and lonely soul that this modern age has spawned by its info-tainment isolation, unless it be in the web chat groups and virtual lives now emerging in the online world of "Second Life."[2] The continuing failure of television programming to become stable is an indication of social mechanisms failing to find healthy activity for our idle time. One possible solution has been working longer hours.

Society has not adequately addresssed these technology-shifted elements of social interaction and needs to. No more summer porch conversations between neighbors. This unmet need continues and I see no solutions emerging unless society itself acts. Science is not going to save us in this particular arena.

Finally in regard to the web, the nature of news will change. With the easy access to web-distributed information via RSS[3] and other tools that support blogs, the Associated Press and Reuters may reconstitute themselves into *credibility* rating services for the many different sources of news rather than being sources themselves.

The Touch-Store

Some stores will continue as they have for many centuries. One example is the grocery store, where sight and feel are immediate and necessary to enable a sale. Similarly with books, though best sellers may exit the libraries and move to web sellers, finding a reference or non-fiction book or a new novel that suits one's special appetite will continue to require the individual physical sampling provided by a bookstore and library. Web-banks and machine-tellers, on the other hand, have already changed the nature of in-person banks, whose branches have become smaller and with more personal service.

It does seem clear that the web renter/seller of DVD movies will prosper since taste in movies is so individual. No one can stock enough titles in a touch-store to provide the age appropriate range for 80 years of users, never mind make the collection easy to search. Getting on the web and ordering specifically what you want is likely to kill the touch store market for DVDs.

Touch-stores will provide specialized clothing where feel, look and fitting are critical. Malls will become social spaces and places for the display of art in order to retain their comfortable attraction compared to the web purchase of one's personal and predictable size and quality of underwear.

Advertising — an excessive influence?

This 21st Century will drown us in visual advertising. Management of its intrusion and influence has become the basis of several recent novels (*Pattern Recognition* by Neal Stephenson; *Idoru* by William Gibson) attempting to make predictions of our advertising future. There are one hundred plus channels of television, none of them free of advertising. Stadiums are covered with corporate brand names, all manner of visual information inundate us with advertising for products. Because of the money behind advertising we are better advised to regulate advertising than to regulate the cloning mentioned earlier.

My view is that the 21st Century's new capitalist conundrum is how to gain humanitarian control of product marketing, whose financial resources and consequences are nearly equal to federal budget allocations for major projects. A new bill of rights may have to be established to regulate the effects and power of advertising.

World Economics

I make one simple point on this complex front of worldwide expansion. At present we depend on a mix of natural market mechanisms and national self-interest to regulate the world's economy. We are sorely in need of a John Maynard Keynes to establish a new paradigm for forging the disparate national economies into a fair and proportionally successful system. Sadly no such champion has emerged. That's it. That is my futurist limited telescope image from the past. The WTO is our present best (and beleaguered) effort. Countries and continents will consolidate and attempt to establish balance of their national resources — physical labor, intellectual labor, hardware manufacturing, in-person and remote service provision, and all the forms of selling, in terms of the world demand. These national resources will be required for the same manufacturing commodities and meeting of human needs that presently drive the market: food and clothing, energy supplies, health services and medication, transportation, entertainment, and finally police and military protection.

Another 21st Century economic translocation of labor is the use of remotely located low cost labor to provide customer support and information management. This trend, called outsourcing, will broaden the cash flow from industrial and design intensive nations toward the economically emerging nations who show interest in providing such support and who can also provide the educational infrastructure needed to prepare their populations for the work. The management of information may become the largest source of employment in the new century, leaving manufacturing in second place.

Space

Slow advances will occur in the uses and discoveries of interplanetary space unless nuclear war becomes a force for mass migration. Barring scientific breakthroughs in control of gravity or time, the benefits from advances in physics will remain terrestrial. In fact physics may have peaked in the 20th Century just as the novel has.

Energy — Oil costs and availability

Nuclear power will reemerge as a central source of electrical energy with coal continuing to dominate power generation. Oil will rise in price so much that it will be used exclusively for personal automobiles. The airline industry will undergo a major restructuring, possibly disappearing as a major force for mobility and fresh food. Railroads could make a comeback. Shipping will continue using one form of energy or another. Nuclear could even become mainstream for shipping. The Hydrogen car may come into its own late in the century though exactly what form of motive energy will rise when gasoline costs are greater than house rent is very hard to predict. Electric commuter trains powered by nuclear electricity will surely dominate travel in later centuries.

Regulatory Agencies

The growth of government agencies to protect us is interesting and influential. All of industry tends to fall under its purview. The CDC, the FCC, FAA, FTC are all watchdogs over various aspects of human safety and protection. These are the names for those in the United States. Each developed country has its own watchdog agency. Their interlocking has been tested most in the newly formed European Union and possibly that model is a guide for their evolution in the coming times for unifying regulation in the world economy. Trade has forced these countries to adopt uniform standards — perhaps the single most relevant aspect of regulation that will continue to evolve in this century. At some point I see the emergence of a price label indicating the cost of compliance and so providing us with a tool that will help us regulate the regulators.

Continents

Africa will continue to be the slowest developing continent and efforts to alter that will be unsuccessful. Africa has a rich natural resource bed and may come into its own in the 22nd century, once its leaders and populace become more educated and less tribally divided. The Middle East will evolve best after oil runs out and its people have to survive without its easily derived income.

CLOSING

Here is the good news. In spite of the upheavals that seem inevitable, our daily work, our consumption of goods, our thinking and ideas press us forward, helping us gain ground on the past in ways that, looking back, we are proud of.

There are many who look forward at possibilities of catastrophe and they may be our prophets, warning us to use care. There are those who dream of alternatives and build them and test them in the human marketplace for acceptability. There are the many of us who find a niche and live out our lives richly and poorly there.

In the end, as I told my son when he was twelve, the video games will entertain you for a while, but the investigation of life seems, so far, inexhaustible in its rich provision of engaging dissection and discovery.

Reference notes
1. Niel A. Campbell, *Biology,* 4th Edition, (Benjamin/Cummings Publishing Co., Inc. Menlo Park CA , 2000): 303.
2. http://secondlife.com/whatis

3. RSS stands for "Really Simple Syndication" and is web software that allows constant updating of content in a manner similar to news updates and is most effectively used presently as a news source. Anyone can create an RSS feed and become a source of easily updated information to those who seek that particular RSS source.

APPENDICES

APPENDIX 1

LIST BY YEAR

CHANGE EVENTS	YEAR	CHANGE AGENT	CATEGORY
"Serpentine Dance"	1900	Loie Fuller	Fine Art, Music and Dance
Quantum Theory	1900	Max Planck	Knowledge of Universe
Transition Poetry	1900	William Butler Yeats	Poetry and Literature
Interpretation of Dreams	1900	Sigmund Freud	Psychology
Mathematics of Approximation, Computers	1900	Taylor, Hermite, Hamiltonian	Technical Uncertainty
White Rice vs. Brown (health factor)	1901	Christaan Eijkman/Gerrit Grijns	Health
Safety Razor	1901	King Camp Gillette	Life Utilities
Gyroscope (MIT)	1901	Charles Draper	Mobility/Communication
Transatlantic wireless telegraph	1901	Guglielmo Marconi/Karl F. Braun	Mobility/Communication
X-Rays	1901	Wilhelm G. Roentgen	Understanding the Body
Air Conditioning, temperature control etc.	1902	Willis Haviland Carrier	Life Utilities
Black and White Photography as Art	1902	Alfred Stieglitz	Recording Life
Genetics and Inheritance	1902	Mendel/Fleming/De Bries/Sutton	Understanding the Body
"Dance of the Future"	1903	Isadora Duncan	Fine Art, Music and Dance
Invention of the Electric Utility	1903	Thomas Alva Edison	Life Utilities
Powered Flight	1903	Wilbur and Orville Wright	Mobility/Communication
Vacuum Tube, Diode/Rectifier/Detector	1904	Sir John Ambrose Fleming	Mobility/Communication
Offset Printing becomes real	1904	Ira Rubel	Mobility/Communication

1905	Albert Einstein-Based on Planck	Invention of light quanta	Knowledge of Universe
1905	Albert Einstein	Special Relativity E=mC^2	Knowledge of Universe
1905	A. Binet and T. Simon	IQ Test	Psychology
1906	W. Fletcher / F. Hopkins	Importance of Vitamins to health	Health
1906	August von Wassermann	Wassermann Test for Syphilis	Health
1906	Maria Montessori	Montessori Method	Learn and Teach Tools
1906	Upton Sinclair	The Jungle	Poetry and Literature
1906	Finland's Diet	Women's Vote	Social Change
1907	Bertram Borden Boltwood	Radioactive Decay (Carbon) Dating	Earth History
1907	Pablo Picasso	Modern Art (plus hundreds of artists)	Fine Art, Music and Dance
1908	Lee de Forest	RF Amplifier Vacuum Tube, Audion, Triode	Mobility/Communication
1908	Henry Ford	Automobile Assembly Line	Mobility/Communication
1909	Frank Lloyd Wright	American Architecture	Dwellings and Structures
1910	Leo Baekeland	Bakelite (first fully synthetic plastic)	Life Utilities
1910	P. de Vivie/T. Campagnolo/Tour de France	24 Gear Mountain Bikes	Life Utilities
1910	Pieter Zeeman/ Hendrick A. Lorentz	Electro-Magnetic waves	Mobility/Communication
1910	Bertrand Russell	Interesting Problems	Philosophy
1910	China	China Abolishes Slavery	Social Change
1911	C. Funk	Vitamin(e) Chemistry — an Amine	Health
1911	H.K. Onnes	Superconductivity	Knowledge of Universe
1911	Ernest Rutherford	Atomic Nucleus	Knowledge of Universe

CHANGE EVENTS	YEAR	CHANGE AGENT	CATEGORY
Inside the Mind	1911	Alfred Adler	Psychology
Ragtime respectability — Carnegie Hall	1912	James R. Europe, Clef Club Band	Fine Art, Music and Dance
Oil Cracking for Gasoline	1912	William Burton	Mobility/Communication
Sinking of Titanic	1912	SS Titanic	Technical Uncertainty
New York Armory Show of modern art	1913	W. Kuhn, A. B. Davies	Fine Art, Music and Dance
16th Amendment	1913	Personal Income Tax	Government
First Vitamin (A) isolated	1913	Davis and McCollum-University of Wisconsin/ Osborn and Mendel-Yale	Health
Subatomic Behavior	1913	Niels Bohr	Knowledge of Universe
Home Refrigerator (Domestic Electric Refrigerator)	1913	Domelre by F. Wolf /A. Mellowes	Life Utilities
Beginnings of Behaviorist theory	1913	John B. Watson	Psychology
Panama Canal	1914	United States	Dwellings and Structures
Blood Bank	1914	Albert Hustin	Health
Star Lives — The Main Sequence	1914	Henry Russell	Knowledge of Universe
Proton	1914	Ernest Rutherford	Knowledge of Universe
Super Regenerative Radio Amplifier	1914	Edwin Howard Armstong	Mobility/Communication
WWI End of Colonialism	1914	France/Germany	War
Origins of Continents and Oceans	1915	Alfred Lothar Wegener	Earth History
The Trial	1915	Franz Kafka	Government
Chlorination of Water — purification	1915	Abel Wolmand	Health
Movie Comedy	1915	Charlie Chaplin	Recording Life

First American Birth Control Clinic (becomes Planned Parenthood in 1942)	1915	Margaret Sanger	Social Change
Movies: MGM	1917	Louis Meyer	Recording Life
Bolshevik Revolt	1917	Vladimir Lenin	Social Change
First Congresswoman	1917	Jeannette Rankin	Social Change
Twenty to forty million influenza deaths	1918	Swine (Spanish) Flu	Health
Classical Guitar	1920	Andres Segovia	Fine Art, Music and Dance
Tractates Logico — Philosophicus	1920	Ludwig Wittgenstein	Philosophy
19th Amendment — Women's' Vote	1920	Cady Stanton/ Susan B. Anthony	Social Change
Prohibition-Volstead Act+18th Amendment	1920	WCTU/Anti-Saloon League	Social Change
Progressive Civil Disobedience	1920	Mohandas K. Gandhi	Social Change
Music and Medicine for the poor	1920	Albert Schweitzer	Social Change
12-Tone Serial Technique	1921	Arnold Schoenberg	Fine Art, Music and Dance
Insulin for habilitation of diabetics	1922	F. Banting, C. Best; J.J.R. McCleod, J. Collip, University of Toronto	Health
Sound Movies	1922	T.W. Case	Mobility/Communication
The Skyscrapers	1923	Charles E. Jeanneret/Le Corbusier	Dwellings and Structures
Creole Jazz Band	1923	King Oliver, Jelly Roll Morton	Fine Art, Music and Dance
Wave/Particle Duality	1923	Louis de Broglie	Knowledge of Universe
Frozen Fish Process	1923	Clarence Birdseye	Life Utilities
Electro-static Precipitator — removes particulates	1923	Fred Cottrell	Life Utilities
TV Camera Tube (orthicon, vidicon)	1923	Vladimir Kosma Zworykin	Mobility/Communication
Modernized Turkey for Muslim Women	1923	Mustafa Kemal Pasha Ataturk	Social Change

CHANGE EVENTS	YEAR	CHANGE AGENT	CATEGORY
Functional Architecture	1924	Walter Gropius	Dwellings and Structures
Universe Bigger than Milky Way	1924	Edwin Hubbel	Knowledge of Universe
The Loud Speaker	1924	Rich Kellog	Mobility/Communication
Americanization of English humor	1925	James Thurber	Humor
Matrix Mechanics	1925	Werner Heisenberg	Knowledge of Universe
The Hollow Men	1925	T.S. Eliot	Poetry and Literature
New Means of Conveying Comedy	1926	Buster H. Keaton and Harold Lloyd	Humor
Rocket "for Extreme Altitudes" (V2= 1944)	1926	Robert H. Goddard	Mobility/Communication
X-ray CAT Scans	1926	Bob Ledley	Understanding the Body
Failed Marxism policies	1926	Joseph Stalin	War
Father of Blues	1927	Son House	Fine Art, Music and Dance
Uncertainty Physics	1927	Born/Schroedinger/Heisenberg	Knowledge of Universe
PVC (Polyvinyl Chloride) Modern Plastic	1927	Fritz Klatte(1912), Waldo Semon	Life Utilities
Real Sound Movies — The Jazz Singer	1927	Warner Brothers	Mobility/Communication
Negative Feedback Amplifier	1927	Harold Black	Mobility/Communication
Quartz Clock — accurate time keeping	1927	W.A. Marrison and J.W. Horton	Mobility/Communication
Sports Hero Model	1927	Babe Ruth	Social Change
Classical Music — Musical Theater	1928	George Gershwin	Fine Art, Music and Dance
Penicillin discovered	1928	Alexander Fleming	Health
Magnetic Wire Recording (V. Poulsen 1898)	1928	J. Begun/ F. Pfleumer	Mobility/Communication
Color Film	1928	George Eastman	Recording Life

Scotch Tape	1929	Richard Drew/3M	Life Utilities
Revolutionize Road Transportation	1929	Autobahn — Germany	Mobility/Communication
For Whom the Bell Tolls — We are one nation	1929	Ernest Hemingway	Mobility/Communication
Child Development Phases	1929	Jean Piaget	Psychology
US Stock Market Crash-American Depression	1929	Wall Street USA	Social Change
Jazz Solo Guitar	1930	Charlie Christian / Django Reinhardt	Fine Art, Music and Dance
"Lamentation" "Appalachian Spring"(1942)	1930	Martha Graham and Aaron Copland	Fine Art, Music and Dance
Standup TV Comedy	1930	Marx Brothers	Humor
Cyclotron	1930	Ernesto Lawrence	Knowledge of Universe
CFC = Chlorofluorohydrocarbons — noncorrosive	1930	Thomas Midgley Jr.	Life Utilities
First Electrical Computer	1930	Vannevar Bush	Mobility/Communication
American Ideal Family Paintings	1930	Norman Rockwell	Recording Life
Stereophonic Sound (dual sources)	1931	A.D. English	Fine Art, Music and Dance
Math is Uncertain — Gödel's Proof	1931	Kurt Goedel	Technical Uncertainty
Large structure/sculpture, Mobile	1932	Alexander Calder	Fine Art, Music and Dance
External Heart Pacemaker-unsuccessful	1932	A.S. Hyman	Health
Neutron	1932	James Chadwick	Knowledge of Universe
Brave New World	1932	Aldous Huxley	Recording Life
Polaroid Film	1932	Edwin H. Land	Recording Life
Volstead Act repealed — Prohibition ended	1933	Democratic moralism	Government

CHANGE EVENTS	YEAR	CHANGE AGENT	CATEGORY
First Synthetic Vitamin — Vitamin C	1933	Szent-Gyorgyi/WaughandKing/T. Reichistein, Swiss Inst./C. Haworth, UK Birmingham	Health
e Spin	1933	Paul Dirac	Knowledge of Universe
Non-Local Realism	1933	Puldansky, Rosen, Einstein	Knowledge of Universe
Gold Standard for currency abolished	1933	US Government	Life Utilities
Major Construction Projects-Hoover Dam	1933	Stephen Bechtel	Social Change
Entitlement, Welfare, The New Deal	1933	Franklin D. Roosevelt	Social Change
Prohibition Repealed (21st Amendment)	1933	USA people	Social Change
Comedy Slapstick	1934	Three Stooges	Humor
The Rise and Fall of Cultures	1934	Arnold J. Toynbee	Recording Life
Philos. of Science-"Falsifiability"	1934	Karl Popper	Technical Uncertainty
Sulfa Drugs	1935	Gerhard Domagk-IG Farben/P. Gelmo	Health
Gallup Polls- Reliable Polling	1935	George Gallup	Life Utilities
Airlines- Pan American	1935	John Trip	Mobility/Communication
Polling and statistical sampling	1935	George Gallup	Social Change
Alcoholics Anonymous (AA)	1935	Bill Wilson	Social Change
Major builder of The Blues	1936	Robert Johnson	Fine Art, Music and Dance
Plays of American Severity	1936	Eugene O'Niell	Recording Life
The General Theory of Employment, Interest and Money Deficit Economics	1936	John Mayard Keynes	Social Change
Capitalism/Organized Labor	1936	Walter Reuther	Social Change

American Folk Music	1937	Woody Guthrie	Fine Art, Music and Dance
Middle America Realism Painting	1937	Andrew Wyeth	Fine Art, Music and Dance
Computer Programs/Artificial Intelligence	1937	Alan M. Turing	Learn and Teach Tools
Nylon (synthetic fabrics)	1937	Wallace H. Carothers	Life Utilities
Computers/programs/ Artificial Intelligence	1937	Alan M. Turing	Life Utilities
Jazz Vocal Recordings	1938	Ella Fitzgerald	Fine Art, Music and Dance
Ball Point Pen	1938	Laszlo Biro	Life Utilities
Teflon-PTFE Neutral High-temperature plastic	1938	Roy Plunkett	Life Utilities
Skinner Box modifies behavior	1938	Burrhus Frederic Skinner	Psychology
Penicillin produced	1939	H. Florey-E.Chain-Oxford,UK / A. Morey/J. Kane-Pfizer	Health
Fission	1939	Hahn/Meitner; Strassman/Bohr	Knowledge of Universe
Fission Reaction/Nuclear Electric Power	1939	Enrico Fermi	Life Utilities
DDT= Insect/Typhus Disease Control Pesticide	1939	Muller/Ziedler	Life Utilities
Helicopters — Vertical Lift	1939	Igor Sikorsky	Mobility/Communication
Cathode Ray Tube (CRT)	1939	Zworykin, Farnsworth, Dumont	Mobility/Communication
Television	1939	David Sarnoff	Recording Life
WWII triggered	1939	AdolfHitler	War
Nationalism	1939	Wane of Imperialism	War
Magnetic Tape Recording Heads	1941	Marvin Camras	Mobility/Communication
First Jet Engine Aircraft	1941	Sir Frank Whittle	Mobility/Communication
Duck(duct) Tape	1942	Johnson and Johnson Corp.	Life Utilities

CHANGE EVENTS	YEAR	CHANGE AGENT	CATEGORY
Spread Spectrum Idea, later used for cell phone	1942	Hedy Lamarr and George Antheil	Mobility/Communication
L'Etranger (The Stranger)	1942	Albert Camus	Philosophy
Jazz/Blues	1943	Muddy Watters	Fine Art, Music and Dance
Pap Smear Test	1943	G. Papanicolaou, H.Traut-Columbia U.	Health
Scuba Aqua Lung	1943	Jacques Cousteau	Life Utilities
Philosophy of Existentialism	1943	JP Sartre/Albert Camus	Philosophy
Petit Prince — Friendship and Hope Defined	1943	Antoine de Saint Exupéry	Philosophy
Stage Musicals	1943	Oscar Hammerstein	Recording Life
Cortisone	1944	P. Julian DePau U./L. H. Sarette - Merck Inc/E. Kendall	Health
DNA ID'ed as genetic material	1944	Oswald T. Avery	Understanding the Body
Beam Magnetic Resonance	1944	Isidor Rabi	Understanding the Body
War Generals General	1944	George S. Patton	War
Cyclotron for Physics studies	1945	E.M. McMillan/V. Veksler	Knowledge of Universe
Software — Nanosec. Wire	1945	Adm. Grace Hopper	Learn and Teach Tools
Mathematical Encryption/Security (1918)	1945	Enigma, Bletchley Park	Mobility/Communication
Animal Farm	1945	George	Poetry and Literature
Marshall Plan Bretton Woods	1945	George Marshall	War
Atomic Bomb — Hiroshima	1945	Robert Oppenheimer	War
United Nations Founded	1945	Woodrow Wilson	War
Automotive Smog Defined — Los Angeles	1946	R. R. Tucker/Haagen-Smit	Earth History

Computer Architecture	1946	John von Neumann	Learn and Teach Tools
ENIAC Computer	1946	Mauchly/ Eckert /John /J. Presper Jr.	Learn and Teach Tools
Commonsense Baby and Child Care	1946	Benjamin Spock	Social Change
Nuclear Magnetic Resonance	1946	Edward Purcell and Felix Bloch	Understanding the Body
Hologram Laser 3D Images	1947	Dennis Gabon	Knowledge of Universe
Microwave Oven	1947	Percey L. Spencer	Life Utilities
African American Sports	1947	Jackie Robinson	Social Change
Truman Doctrine	1947	Harry S. Truman	War
Classical American Ballet	1948	George Balanchine	Fine Art, Music and Dance
The LP vinyl record	1948	Peter Goldmark	Fine Art, Music and Dance
Historical Nove	1948	James Albert Michener	Learn and Teach Tools
Velcro (idea from a plant) (hook and loop fastener)	1948	George de Mestral	Life Utilities
The Transistor	1948	Brattain, Bardeen, Shockley	Mobility/Communication
Universal Declaration of Human Rights	1948	Eleanor Roosevelt	Social Change
Popular American Folk Music	1949	Pete Seeger	Fine Art, Music and Dance
Information Theory	1949	C.E. Shannon	Knowledge of Universe
1984 — the Novel	1949	George Orwell	Learn and Teach Tools
Leads on Heart Pacemaker (with external battery)	1950	C. W. Lillehei and E. Bakken	Health
Composite Carbon-fiber Materials. Strength-Weight	1950	Rolls Royce/Royal Air Force	Life Utilities
Cashless Society — Credit Card	1950	Diner's Club	Mobility/Communication
Child Development	1950	Erik Erikson	Psychology
External Heart Pacemaker	1951	J. A. Hopps-Banting Institute	Health

CHANGE EVENTS	YEAR	CHANGE AGENT	CATEGORY
Tough Minded Democracy	1951	Winston Churchill	War
Salk Vaccine for Polio	1952	Jonas Salk	Health
External Portable Resuscitator	1952	P. Zoll- Beth Israel Hospital	Health
Sedatives	1952	Robert Wilkins	Life Utilities
Zone Refining of Silicon	1952	William Pfann	Mobility/Communication
East of Eden	1952	John Steinbeck	Poetry and Literature
Power of Positive Thinking	1952	Norman V. Peale	Psychology
USA Interstate Highway System	1953	President Dwight D. Eisenhower	Life Utilities
Color TV	1953	RCA- NTSC and Shadow Mask	Mobility/Communication
Desegregation, One man-One vote	1953	Supreme Court – Chief Justice Earl Warren	Social Change
DNA double helix structure	1953	James Watson/Francis Crick	Understanding the Body
Rock and Roll	1954	Elvis Presley	Fine Art, Music and Dance
Kidney Transplant	1954	Joseph E. Murray	Health
The Lord of the Flies	1954	William Golding	Poetry and Literature
Sub-Four Minute Mile	1954	Roger Bannister	Understanding the Body
Nuclear Submarine — First Strike threat ended	1954	Adm. Hyman Rickover	War
Disneyland	1955	Walt Disney	Fantasy
Tetracycline broad antibiotic	1955	Lloyd Conover	Health
Anti-proton	1955	E. G. Segre/O. Chamberlain	Knowledge of Universe
Dr. Seuss, Green Eggs and Ham	1955	Theodore Seuss Geisel	Learn and Teach Tools

Rock and Soul	1955	Charles E. Anderson "Chuck" Berry	Mobility/Communication
Fiber Optics	1955	N Kapany/B O'Brien	Mobility/Communication
McDonalds — Father of Fast Food	1955	Ray Kroc	Social Change
Beginning of the end to Racism	1955	Rosa Parks	Social Change
Same Day Global Interconnection	1956	Airfreight	Mobility/Communication
Video Tape	1956	Charles Ginsberg/Ray Dolby	Mobility/Communication
Container Ship — Ideal-X	1956	Malcom Mclean	Mobility/Communication
Int'l. Geophysical Year — atmospheric. monitoring	1957	World governments	Earth History
Interferon (the atom of biology)	1957	A. Isaacs and J. Lindenmann	Health
Superconductivity	1957	J. Bardeen/L.Cooper/J.Schrieffer/U ill.	Knowledge of Universe
Radio Astronomy	1957	Bernard Lovell	Knowledge of Universe
1st Satellite — Sputnik	1957	USSR	Knowledge of Universe
Individual Objectivism	1957	Ayn Rand	Philosophy
Syntactic Structures	1957	Noam Chomsky	Psychology
Implantable Pacemaker	1958	W. Greatbatch and W. Chardack	Health
Xerox Dry Copying	1958	Chester Carlson	Mobility/Communication
Laser/Maser	1958	Gould, Townes and Schawlow	Mobility/Communication
Marxism on Chinese Social Order	1958	Mao Zedong	Social Change
First Rocket reaches the Moon	1959	USSR	Knowledge of Universe
Integrated Circuit	1959	Robert Noyce/Jack Kilby	Mobility/Communication
Aswan High Dam	1960	Egypt	Dwellings and Structures

CHANGE EVENTS	YEAR	CHANGE AGENT	CATEGORY
Ocean Floor Topography — Plate Tectonics	1960	W. Maurice Ewing and Columbia U. team	Earth History
Chimpanzee Tool Use Similar to Human	1960	Jane Goodall	Earth History
Ruby Laser	1960	Theodore Maiman Bloembergen	Mobility/Communication
American Slash Poetry	1960	Anne Sexton	Poetry and Literature
Power to Overcome Incapacity	1960	Helen Keller	Social Change
OPEC Oil Price Control	1960	OPEC	Social Change
Oral Contraceptive (1923-C. Djerassi)	1960	G. Pincus/Min-Chueh Chang/J. Rock	Social Change
Australopithecus — Homo Habilus	1961	Louis Seymour Bazett Leakey	Earth History
1st Personed Satellite	1961	USSR, Gagarin	Mobility/Communication
To Kill a Mockingbird	1961	Harper Lee	Poetry and Literature
Peace Corps of America	1961	John F. Kennedy	Social Change
Genetic Code-DNA=>RNA=Protein	1961	Marshall W. Nirenberg/Gobind Khorana	Understanding the Body
Silent Spring	1962	Rachel Carson	Earth History
LaserandLight Emitting Diode (LED)	1962	Robert Hall/ Nick Holonyak Jr.	Mobility/Communication
First Communications Satellite Telstar I	1962	AT&T/NASA/Bell Labs	Mobility/Communication
Rock and Roll World Wide	1963	The Beatles	Mobility/Communication
Folk Music Re-Invented	1963	Bob Dylan	Social Change
The Feminine Mystique	1963	Betty Frieden	Social Change
LSD — Chemically Altered Perception	1963	Timothy Leary	Social Change
"I Have a Dream "Civil Rights	1963	Martin Luther King Jr.	War

Medicare Act of 1964	1964	Lyndon Baines Johnson	Health
Cosmic Background Radiation	1964	G. Gamow (1916 Dicke)	Knowledge of Universe
High Temperature Superconductors	1964	K.A. Müller/J.D. Bednorz	Knowledge of Universe
Economic Philosophy/ Knowledge Worker	1964	Peter Drucker	Social Change
Black Power Awareness	1964	Malcom X	Social Change
Real Time Ultrasound Tissue Scanning	1964	Siemens	Understanding the Body
Vietnam War — Guerilla War	1964	France, USA	War
Peaceful Protest, UC Berkeley	1964	Mario Savio	War
Gulf of Tonkin Resolution	1964	US Congress	War
India Tonality Influence on Rock via Beatles	1965	Ravi Shankar	Fine Art, Music and Dance
Opera	1965	Luciano Pavarotti	Fine Art, Music and Dance
Miniskirt	1965	Mary Quant	Social Change
First Artificial Heart (short lived success)	1966	Michael Ellis De Bakey	Health
Social Science — Fiction TV Philosopher	1966	Gene Roddenberry	Social Change
Acid Rain	1967	Coal Electric Power Generation	Earth History
Heart Transplant	1967	Christian Barnard	Health
Hippies Summer of Love	1967	Haight Ashbury	Social Change
Green Revolution — higher wheat grain yields	1968	Norman Ernest Borlaug	Earth History
Quarks	1968	H.Freidman/H. Kendall/ R. Taylor	Knowledge of Universe
Electronic Quartz Watch	1968	Centre Electronique Horloge	Life Utilities
Fiber Optic Cable/Communication	1968	Corning Corp	Mobility/Communication
2001: A Space Odyssey	1968	Arthur C. Clark	Poetry and Literature

CHANGE EVENTS	YEAR	CHANGE AGENT	CATEGORY
Child/Adult Moral Development	1968	Lawrence Kohlberg	Psychology
Revisions to Spirituality	1968	Eastern Religions	Social Change
Artificial Heart (partial success)	1969	R. S. Jarvik and P. Winchell	Health
First Person on the Moon	1969	USA Kennedy	Knowledge of Universe
Muppets — Puppet Educators	1969	Jim Henson	Learn and Teach Tools
Powerful Woman Politician	1969	Golda Meir	Social Change
Cleaner Exhaust Automobiles	1970	U.S. Government, EPA Clean Air Act	Earth History
Fiber Optic Telephony (wide use 1977)	1970	AT&T	Mobility/Communication
Public Keys, Encryption	1970	James Ellis, C Clocks-73/Martin Hellman/M Williamson	Mobility/Communication
Feminism	1970	Germain Greer	Social Change
Telethon Fund Raising for the Sick	1970	Jerry Lewis	Social Change
Economics — Government Monetary Policy	1970	Paul A. Samuelson	Social Change
Cloned Living Frog	1970	John Gurdon	Understanding the Body
Recombinant DNA — Restriction Enzyme	1970	H. Smith and K. Wilco	Understanding the Body
First Rocket Landing on Mars	1971	USSR	Knowledge of Universe
Liquid Crystal Display LCD	1971	RCA, Kent State U, LaRoche Corp.	Life Utilities
Pocket Calculator	1971	Texas Instruments Inc.	Life Utilities
"All in the Family" real people	1971	Norman Lear	Recording Life
Imagine — Music of the Heart	1971	John Lennon	War
Kids TV	1973	Mr. Rogers/Jim Henson	Learn and Teach Tools

Roe vs. Wade — Womens Rights	1973	US Supreme Court	Social Change
Imaging-Magnetic Resonance UNY	1973	Paul Lauterbur	Understanding the Body
Ozone Layer — Impact of Freon	1974	Molina, Crutzen,Rowland-UCI	Earth History
Genocide	1975	Pol Pot	War
RSA Public Key (Diffie-Hellman)	1976	Whitfield Diffie/Martin Hellman	Mobility/Communication
Cell Phone	1977	Cooper at Motorola, AT&T	Mobility/Communication
Apple II	1977	Steve Wozniak, Steve Jobs	Mobility/Communication
Future Revealing Movies	1977	George Lucas	Social Change
Cheap High Quality CD Music Reproduction	1979	Joop Sinjou and Toshitada Doi	Mobility/Communication
Dedication to the Helpless	1979	Mother Theresa	Social Change
Government Regulation of Economy	1979	Paul Volker	Social Change
Atomic Power Paranoia	1979	Fission Failure at Three Mile Island	Technical Uncertainty
CAT Scanning	1979	Cornack Hounsfield	Understanding the Body
Post-it Notes	1980	3M corporation — S. Silver/A.Fry	Life Utilities
"Hill Street Blues" real TV drama	1981	Steven Bochco	Recording Life
First Female Supreme Court Justice	1981	Sandra Day O'Connor	Social Change
CAD (Computer aided design) software	1982	Many contibutors (see footnotes)	Life Utilities
Future Revealing Movies	1982	Stephen Spielberg	Social Change
HIV and AIDS Epidemic Scare	1983	Gene Mutation	Social Change
Apple Macintosh Personal Computer	1984	Steve Jobs, Steve Wozniak	Learn and Teach Tools
Documented Ozone Hole in Antarctic	1985	Farman/ Gardiner/Shanklin/UK	Earth History
Fast Mail	1985	Federal Express Co.	Mobility/Communication

CHANGE EVENTS	YEAR	CHANGE AGENT	CATEGORY
Nuclear Reactor disaster — Chernobyl	1986	USSR	Life Utilities
Montreal Protocol — Limit Atmospheric CFCs	1987	United Nations and scientists	Earth History
Global Positioning System	1987	J. J. Spilkher	Mobility/Communication
Revisions to Spirituality	1988	Babba Ram Daas-Richard Alpert	Social Change
World Wide Web (WWW)	1989	DARPA, UCLA, CERN, MIT-Mosaic, HTTP, HTML, TCP/IP, DNS	Mobility/Communication
Muslim Reintegration	1989	Salmon Rushdie	Social Change
Financial Failure of Marxism	1989	USSR Mikhail Gorbachev	Social Change
Easy Graphical Transmission	1990	FAX	Mobility/Communication
Self-help Book Mania	1990	USA boomers Born 1946-64	Psychology
A Brief History of Time	1991	Stephen Hawking	Poetry and Literature
Human Genome Project — Celera Corporation	1991	J. Craig Venter-Celera/ NIH	Understanding the Body
Integration begins in South Africa	1994	Nelson Mandela	Social Change
Kyoto Protocol — Limit CO_2 — Greenhouse	1996	UNFCCC United Nations	Earth History
AIDS Cocktail (to slow the effects)	1996	NIH-many contributors	Health
Over-Population	1999	India	Social Change

APPENDIX 2

LIST BY CATEGORY

CHANGE EVENTS	YEAR	CHANGE AGENT	CATEGORY
American Architecture	1909	Frank Lloyd Wright	Dwellings and Structures
Panama Canal	1914	United States	Dwellings and Structures
The Skyscrapers	1923	Charles E. Jeanneret/Le Corbusier	Dwellings and Structures
Functional Architecture	1924	Walter Gropius	Dwellings and Structures
Aswan High Dam	1960	Egypt	Dwellings and Structures
Radioactive Decay (Carbon) Dating	1907	Bertram Borden Boltwood	Earth History
Origins of Continents and Oceans	1915	Alfred Lothar Wegener	Earth History
Automotive Smog Defined — Los Angeles	1946	R. R. Tucker/Haagen-Smit	Earth History
Int'l. Geophysical Year — atmospheric monitoring	1957	World governments	Earth History
Chimpanzee Tool Use Similar to Human	1960	Jane Goodall	Earth History
Ocean floor topography — Plate Tectonics	1960	W. Maurice Ewing and Columbia	Earth History
Australopithecus — Homo Habilus	1961	Louis Seymour Bazett Leakey	Earth History
Silent Spring	1962	Rachel Carson	Earth History
Acid Rain	1967	Coal Electric Power Generation	Earth History
Green Revolution (higher wheat grain yields)	1968	Norman Ernest Borlaug	Earth History
Cleaner Exhaust Automobiles	1970	US Government, EPA Clean Air Act	Earth History
Ozone Layer Impact of Freon	1974	Molina, Crutzen, Rowland: UCI	Earth History
Documented Ozone Hole in Antarctic	1985	Farman/ Gardiner/Shanklin/UK	Earth History

Montreal Protocol — Limit Atmospheric CFC's	1987 United Nations and scientists	Earth History
Kyoto Protocol — Limit CO2 — Greenhouse	1996 UNFCCC	Earth History
Disneyland	1955 Walt Disney	Fantasy
"Serpentine Dance"	1900 Loie Fuller	Fine Art, Music and Dance
"Dance of the Future"	1903 Isadora Duncan	Fine Art, Music and Dance
Modern Art (along with hundreds of other artists)	1907 Pablo Picasso	Fine Art, Music and Dance
Ragtime respectability — Carnegie Hall	1912 James R. Europe, Clef Club Band	Fine Art, Music and Dance
New York Armory Show of modern art	1913 W. Kuhn, A. B. Davies	Fine Art, Music and Dance
Classical Guitar	1920 Andres Segovia	Fine Art, Music and Dance
12-Tone Serial Technique	1921 Arnold Schoenberg	Fine Art, Music and Dance
Creole Jazz Band	1923 King Oliver, Jelly Roll Morton	Fine Art, Music and Dance
Father of Blues	1927 Son House	Fine Art, Music and Dance
Classical Music — Musical Drama	1928 George Gershwin	Fine Art, Music and Dance
"Lamentation" "Appalachian Spring"(1942)	1930 Martha Graham and Aaron Copland	Fine Art, Music and Dance
Stereophonic Sound (dual sources)	1931 A.D. English	Fine Art, Music and Dance
Large Structure/Sculpture, Mobile	1932 Alexander Calder	Fine Art, Music and Dance
Jazz Solo Guitar	1933 Charlie Christian/Django Reinhardt	Fine Art, Music and Dance
Major Builder of The Blues	1936 Robert Johnson	Fine Art, Music and Dance
Middle America Realism Painting	1937 Andrew Wyeth	Fine Art, Music and Dance
American Folk Music	1937 Woody Guthrie	Fine Art, Music and Dance
Jazz Vocal Recordings	1938 Ella Fitzgerald	Fine Art, Music and Dance
Jazz/Blues	1943 Muddy Watters	Fine Art, Music and Dance

CHANGE EVENTS	YEAR	CHANGE AGENT	CATEGORY
Classical American Ballet	1948	George Balanchine	Fine Art, Music and Dance
The LP Vinyl Record	1948	Peter Goldmark	Fine Art, Music and Dance
Popular American Folk Music	1949	Pete Seeger	Fine Art, Music and Dance
Rock and Roll	1954	Elvis Presley	Fine Art, Music and Dance
India Tonality Influence on Rock via The Beatles	1956	Ravi Shankar	Fine Art, Music and Dance
Opera	1965	Luciano Pavarotti	Fine Art, Music and Dance
16th Amendment	1913	Personal Income Tax	Government
The Trial	1915	Franz Kafka	Government
Volstead Act repealed — Prohibition ended	1933	Democratic moralism	Government
White Rice vs. Brown — health factor	1901	Christaan Eijkman/Gerrit Grijns	Health
Wassermann Test for Syphilis	1906	August von Wassermann	Health
Importance of Vitamins to Health	1906	W. Fletcher / F. Hopkins	Health
Vitamin(e) Chemistry — an Amine	1911	C. Funk	Health
First Vitamin (A) Isolated	1913	Davis and McCollum-U. Wisc./ Osborn and Mendel-Yale Univ.	Health
Blood Bank	1914	Albert Hustin	Health
Chlorination of Water — purification	1915	Abel Wolmand	Health
Twenty to Forty Million Influenza Deaths	1918	Swine (Spanish) Flu	Health
Insulin for Habilitation of Diabetics	1922	F. Banting, C. Best; J.J.R. McCleod, J. Collip: University of Toronto	Health
Penicillin Discovered	1928	Alexander Fleming	Health
External Heart Pacemaker — unsuccessful	1932	A.S. Hyman	Health

First Synthetic Vitamin — Vitamin C	1933	Szent-Gyorgyi/WaughandKing/T. Reichistein: Swiss Inst. Tech./C. Haworth: UK Birmingham	Health
Sulfa Drugs	1935	Gerhard Domagk-IG Farben/P. Gelmo	Health
Pap Smear Test	1943	G. Papanicolaou, H.Traut:Columbia U.	Health
Cortisone	1944	P. Julian DePau U./L. H. Sarette - Merck Inc/E. Kendall	Health
External Heart Pacemaker	1951	J. A. Hopps, Banting Institute	Health
Salk Vaccine for Polio	1952	Jonas Salk	Health
External Portable Resuscitator	1952	P. Zoll, Beth Israel Hospital	Health
Kidney Transplant	1954	Joseph E. Murray	Health
Leads on Heart Pacemaker with External Battery	1955	C. W. Lillehei and E. Bakken	Health
Tetracycline (broad antibiotic)	1955	Lloyd Conover	Health
Interferon (the atom of biology)	1957	A. Isaacs and J. Lindenmann	Health
Implantable Pacemaker	1958	W. Greatbatch and W. Chardack	Health
Medicare Act of 1964	1964	Lyndon Baines Johnson	Health
First Artificial Heart (short lived success)	1966	Michael Ellis De Bakey	Health
Heart Transplant	1967	Christian Barnard	Health
Artificial Heart (partial success)	1969	R. S. Jarvik and P. Winchell	Health
AIDS Cocktail to slow the effects	1996	NIH-many contributors	Health
Penicillin Produced	1939-1945	H. Florey-E.Chain: Oxford,UK /A. Morey/J. Kane:Pfizer	Health
Americanization of English Humor	1925	James Thurber	Humor

CHANGE EVENTS	YEAR	CHANGE AGENT	CATEGORY
New Means of Conveying Comedy	1926	Buster H. Keaton and Harold Lloyd	Humor
Standup TV Comedy	1930	Marx Brothers	Humor
Comedy Slapstick	1934	Three Stooges	Humor
Quantum Theory	1900	Max Planck	Knowledge of the Universe
Special Relativity E=mC²	1905	Albert Einstein	Knowledge of the Universe
Invention of Light Quanta	1905	Albert Einstein-Based on Planck	Knowledge of the Universe
Atomic Nucleus	1911	Ernest Rutherford	Knowledge of the Universe
Superconductivity	1911	H.K. Onnes	Knowledge of the Universe
Subatomic Behavior	1913	Niels Bohr	Knowledge of the Universe
Proton	1914	Ernest Rutherford	Knowledge of the Universe
Star Lives — The Main Sequence	1914	Henry Russell	Knowledge of the Universe
Wave/Particle Duality	1923	Louis de Broglie	Knowledge of the Universe
Universe Bigger than Milky Way	1924	Edwin Hubbel	Knowledge of the Universe
Matrix Mechanics	1925	Werner Heisenberg	Knowledge of the Universe
Uncertainty Physics	1927	Born/Schroedinger/Heisenberg	Knowledge of the Universe
Cyclotron	1930	Ernesto Lawrence	Knowledge of the Universe
Neutron	1932	James Chadwick	Knowledge of the Universe
e Spin	1933	Paul Dirac	Knowledge of the Universe
Non-Local Realism	1933	Puldansky, Rosen, Einstein	Knowledge of the Universe
Fission	1939	Hahn/Meitner; Strassman/Bohr	Knowledge of the Universe
Cyclotron for Physics Studies	1945	E.M. McMillan/V. Veksler	Knowledge of the Universe

Hologram Laser 3D Images	1947	Dennis Gabon	Knowledge of the Universe
Information Theory	1949	C.E. Shannon	Knowledge of the Universe
Anti-proton	1955	E. G. Segre/O. Chamberlain	Knowledge of the Universe
Radio Astronomy	1957	Bernard Lovell	Knowledge of the Universe
Superconductivity	1957	J. Bardeen/L.Cooper/J.Schrieffer/Ul	Knowledge of the Universe
First Satellite — Sputnik	1957	USSR	Knowledge of the Universe
First Rocket Reaches the Moon	1959	USA	Knowledge of the Universe
Cosmic Background Radiation	1964	G. Gamow (1916 Dicke)	Knowledge of the Universe
High Temperature Superconductors	1964	K.A. Müller/J.D. Bednorz	Knowledge of the Universe
Quarks	1968	H.Freidman/H. Kendall/ R. Taylor	Knowledge of the Universe
First Person on the Moon	1969	USA Kennedy	Knowledge of the Universe
First Rocket Landing on Mars	1971	USSR	Knowledge of the Universe
Montessori Method	1906	Maria Montessori	Learn and Teach, New Tools
Computer Programs/Artificial Intelligence	1937	Alan M. Turing	Learn and Teach, New Tools
Software — Nanosecond Wire	1945	Adm. Grace Hopper	Learn and Teach, New Tools
Computer Architecture	1946	John von Neumann	Learn and Teach, New Tools
ENIAC Computer	1946	Mauchly/ Eckert /John /J. Presper Jr.	Learn and Teach, New Tools
Historical Novel	1948	James Albert Michener	Learn and Teach, New Tools
1984, Brave New World	1949	George Orwell/Aldous Huxley	Learn and Teach, New Tools
Dr. Seuss, Green Eggs and Ham	1955	Theodore Seuss Geisel	Learn and Teach, New Tools
Muppets — Puppet Educators	1969	Jim Henson	Learn and Teach, New Tools

CHANGE EVENTS	YEAR	CHANGE AGENT	CATEGORY
Kids TV	1973	Mr. Rogers/Jim Henson	Learn and Teach, New Tools
Apple Macintosh Personal Computer	1984	Steve jobs Steve Wozniak	Learn and Teach, New Tools
Safety Razor	1901	King Camp Gillette	Life Utilities
Air Conditioning, temperature control etc.	1902	Willis Haviland Carrier	Life Utilities
Invention of the Electric Utility	1903	Thomas Alva Edison	Life Utilities
Bakelite (first fully synthetic plastic)	1910	Leo Baekeland	Life Utilities
Home Refrigerator	1913	Domelre by F. Wolf /A. Mellowes	Life Utilities
Frozen Fish Process	1923	Clarence Birdseye	Life Utilities
Electro-static Precipitator (removes particulates)	1923	Fred Cottrell	Life Utilities
PVC (Polyvinyl Chloride) Modern Plastic	1927	Fritz Klatte(1912), Waldo Semon	Life Utilities
Scotch Tape	1929	Richard Drew/3M	Life Utilities
CFC = Chlorofluorohydrocarbons, noncorrosive fluid	1930	Thomas Midgley Jr.	Life Utilities
Gold Standard for currency abolished	1933	US Government	Life Utilities
Computers/programs/ Artificial Intelligence	1937	Alan M. Turing	Life Utilities
Nylon (synthetic fabrics)	1937	Wallace H. Carothers	Life Utilities
Ball Point Pen	1938	Laszlo Biro	Life Utilities
Teflon — PTFE Neutral High-Temperature Plastic	1938	Roy Plunkett	Life Utilities
Fission Reaction/Nuclear Electric Power	1939	Enrico Fermi	Life Utilities
DDT= Insect/Typhus disease control pesticide	1939	Muller/Ziedler	Life Utilities
Duck(duct) Tape	1942	Johnson and Johnson Corp.	Life Utilities

Scuba Aqua Lung	1943	Jacques Cousteau	Life Utilities
Microwave Oven	1947	Percey L. Spencer	Life Utilities
Velcro (idea from a plant)(hook and loop fastener)	1948	George de Mestral	Life Utilities
Sedatives	1952	Robert Wilkins	Life Utilities
USA Interstate Highway System	1953	President Dwight D. Eisenhower	Life Utilities
Electronic Quartz Watch	1968	Centre Electronique Horloge	Life Utilities
Liquid Crystal Display LCD	1971	RCA, Kent State Univ., LaRoche Corp.	Life Utilities
Pocket Calculator	1971	Texas Instruments Inc.	Life Utilities
Post-it Notes	1980	3M corp.–S. Silver/A.Fry	Life Utilities
CAD (Computer aided design) software	1982	Many contibutors (see footnotes)	Life Utilities
Nuclear Reactor disaster — Chernobyl	1986	USSR	Life Utilities
24-gear Mountain Bikes	1910-1987	P. de Vivie/T. Campagnolo/Tour de France	Life Utilities
Composite Carbon-fiber Materials. Strength-Weight	1950s	Rolls Royce/Royal Air Force	Life Utilities
Gyroscope (MIT)	1901	Charles Draper	Mobility/Communication
Transatlantic wireless telegraph	1901	Guglielmo Marconi/Karl F. Braun	Mobility/Communication
Powered Flight	1903	Wilbur and Orville Wright	Mobility/Communication
Offset Printing becomes real	1904	Ira Rubel	Mobility/Communication
Vacuum Tube — Diode/Rectifier/Detector	1904	Sir John Ambrose Fleming	Mobility/Communication
Automobile Assembly Line	1908	Henry Ford	Mobility/Communication
RF Amplifier Vacuum Tube, Audion, Triode	1908	Lee de Forest	Mobility/Communication

CHANGE EVENTS	YEAR	CHANGE AGENT	CATEGORY
Electro-Magnetic Waves	1910	Pieter Zeeman/ Hendrick A. Lorentz	Mobility/Communication
Oil Cracking for Gasoline	1912	William Burton	Mobility/Communication
Super Regenerative Radio Amplifier	1914	Edwin Howard Armstong	Mobility/Communication
Sound Movies	1922	T.W. Case	Mobility/Communication
TV Camera Tube (orthicon, vidicon)	1923	Vladimir Kosma Zworykin	Mobility/Communication
The Loud Speaker	1924	Rich Kellog	Mobility/Communication
"Rocket Reaching Extreme Altitudes" (V2= 1944)	1926	Robert H. Goddard	Mobility/Communication
Negative Feedback Amplifier	1927	Harold Black	Mobility/Communication
Quartz Clock (accurate time keeping)	1927	W.A. Marrison and J.W. Horton	Mobility/Communication
Real Sound Movies — *The Jazz Singer*	1927	Warner Brothers	Mobility/Communication
Magnetic Wire Recording (V. Poulsen 1898)	1928	J. Begun/ F. Pfleumer	Mobility/Communication
For Whom the Bell Tolls — We are one nation	1929	Ernest Hemingway	Mobility/Communication
Revolutionize Road Transportation	1929	Autobahn — Germany	Mobility/Communication
First Electrical Computer	1930	Vannevar Bush	Mobility/Communication
Gallup Polls — Reliable Polling	1935	George Gallup	Mobility/Communication
Airlines — Pan American	1935	John Trip	Mobility/Communication
Helicopters — Vertical Lift	1939	Igor Sikorsky	Mobility/Communication
Cathode Ray Tube (CRT)	1939	Zworykin, Farnsworth, Dumont	Mobility/Communication
Magnetic Tape Recording Heads	1941	Marvin Camras	Mobility/Communication
First Jet Engine Aircraft	1941	Sir Frank Whittle	Mobility/Communication
Spread Spectrum idea, later used for cell phone	1942	Hedy Lamarr and George Antheil	Mobility/Communication

Mathematical Encryption/Security (1918)	1945	Enigma, Bletchley Park	Mobility/Communication
The Transistor	1948	Brattain, Bardeen, Shockley	Mobility/Communication
Cashless Society — Credit Card	1950	Diner's Club	Mobility/Communication
Zone Refining of Silicon	1952	William Pfann	Mobility/Communication
Color TV	1953	RCA- NTSC and Shadow Mask	Mobility/Communication
Rock and Soul	1955	Charles E. Anderson "Chuck" Berry	Mobility/Communication
Fiber Optics	1955	N Kapany/B O'Brien	Mobility/Communication
Same Day Global Interconnection	1956	Airfreight	Mobility/Communication
Video Tape	1956	Charles Ginsberg/Ray Dolby	Mobility/Communication
Container Ship — Ideal-X	1956	Malcom Mclean	Mobility/Communication
Xerox dry Copying	1958	Chester Carlson	Mobility/Communication
Laser/Maser	1958	Gould, Townes and Schawlow	Mobility/Communication
Integrated Circuit	1959	Robert Noyce/Jack Kilby	Mobility/Communication
Ruby Laser	1960	Theodore Maiman Bloembergen	Mobility/Communication
First Personed Satellite	1961	USSR, Gagarin	Mobility/Communication
First Communications Satellite Telstar I	1962	AT&T/NASA/Bell Labs	Mobility/Communication
LaserandLight Emitting Diode-LED	1962	Robert Hall/ Nick Holonyak Jr.	Mobility/Communication
Rock and Roll world wide	1963	The Beatles	Mobility/Communication
Fiber Optic Cable/Communication	1968	Corning Corp	Mobility/Communication
Fiber Optic telephony (wide use 1977)	1970	AT&T	Mobility/Communication
Public Keys, Encryption	1970	James Ellis, C Clocks-73/Martin Hellman/M Williamson	Mobility/Communication

CHANGE EVENTS	YEAR	CHANGE AGENT	CATEGORY
RSA Public Key (Diffie-Hellman)	1976	Whitfield Diffie/Martin Hellman	Mobility/Communication
Cell Phone	1977	Cooper at Motorola, AT&T	Mobility/Communication
Apple II	1977	Steve Wozniak, Steve Jobs	Mobility/Communication
Cheap High Quality CD Music Reproduction	1979	Joop Sinjou and Toshitada Doi	Mobility/Communication
Fast Mail	1985	Federal Express Co.	Mobility/Communication
Global Positioning System	1987	J. J. Spilkher	Mobility/Communication
World Wide Web (WWW)	1989	DARPA, UCLA, CERN, MIT-Mosaic, HTTP, HTML, TCP/IP, DNS	Mobility/Communication
Easy Graphical Transmission	1990	FAX	Mobility/Communication
Interesting Problems	1910	Bertrand Russell	Philosophy
Tractates Logico-Philosophicus	1920	Ludwig Wittgenstein	Philosophy
L' Etranger (The Stranger)	1942	Albert Camus	Philosophy
Petit Prince-Friendship and Hope Defined	1943	Antoine de Saint Exupéry	Philosophy
Philosophy of Existentialism	1943	JP Sartre/Albert Camus	Philosophy
Objectivism — Individual	1957	Ayn Rand	Philosophy
Transition Poetry	1900	William Butler Yeats	Poetry and Literature
Offset Printing becomes real	1904	Ira Rubel	Poetry and Literature
The Jungle	1906	Upton Sinclair	Poetry and Literature
The Hollow Men	1925	T.S. Eliot	Poetry and Literature
Brave New World	1932	Aldous Huxley	Poetry and Literature
Animal Farm	1945	George Orwell	Poetry and Literature

East of Eden	1952	John Steinbeck	Poetry and Literature
The Lord of the Flies	1954	William Golding	Poetry and Literature
American slash poetry	1960	Anne Sexton	Poetry and Literature
To Kill a Mockingbird	1961	Harper Lee	Poetry and Literature
2001: A Space Odyssey	1968	Arthur C. Clark	Poetry and Literature
A Brief History of Time	1991	Stephen Hawking	Poetry and Literature
Interpretation of Dreams	1900	Sigmund Freud	Psychology
IQ Test	1905	A. Binet and T. Simon	Psychology
Inside the Mind	1911	Alfred Adler	Psychology
Beginnings of Behaviorist theory	1913	John B. Watson	Psychology
Child Development Phases	1929	Jean Piaget	Psychology
Skinner Box Modifies Behavior	1938	Burrhus Frederic Skinner	Psychology
Child Development	1950	Erik Erikson	Psychology
Power of Positive Thinking	1952	Norman V. Peale	Psychology
Syntactic Structures	1957	Noam Chomsky	Psychology
Child/Adult Moral Development	1968	Lawrence Kohlberg	Psychology
Self-help Book Mania	1990	USA boomers born 1946-64	Psychology
Black and White Photography as Art	1902	Alfred Stieglitz	Recording Life
Movie Comedy	1915	Charlie Chaplin	Recording Life
Movies: MGM	1917	Louis Meyer	Recording Life
Color Film	1928	George Eastman	Recording Life
American Ideal Family Paintings	1930	Norman Rockwell	Recording Life

CHANGE EVENTS	YEAR	CHANGE AGENT	CATEGORY
Brave New World	1932	Aldous Huxley	Recording Life
Polaroid Film	1932	Edwin H. Land	Recording Life
The Rise and Fall of Cultures	1934	Arnold J. Toynbee	Recording Life
Plays of American Severity	1936	Eugene O'Niell	Recording Life
Television	1939	David Sarnoff	Recording Life
Stage Musicals	1943	Oscar Hammerstein	Recording Life
"All in the Family" real people	1971	Norman Lear	Recording Life
"Hill Street Blues" real TV drama	1981	Steven Bochco	Recording Life
Women's Vote	1906	Finland's Diet	Social Change
China Abolishes Slavery	1910	China	Social Change
First American Birth Control Clinic (Becomes Planned Parenthood in 1942)	1915	Margaret Sanger	Social Change
First Congresswoman	1917	Jeannette Rankin	Social Change
Bolshevik Revolt	1917	Vladimir Lenin	Social Change
Music and Medicine for the poor	1920	Albert Schweitzer	Social Change
19th Amendment — Women's Vote	1920	Cady Stanton/ Susan B. Anthony	Social Change
Progressive Civil Disobedience	1920	Mohandas K. Gandhi	Social Change
Prohibition — Volstead Act+18th Amendment	1920	WCTU/Anti-Saloon League	Social Change
Modernized Turkey for Muslim Women	1923	Mustafa Kemal Pasha Ataturk	Social Change
Sports Hero Model	1927	Babe Ruth	Social Change
US Stock Market Crash — American Depression	1929	Wall Street USA	Social Change

Entitlement, Welfare, The New Deal	1933	Franklin D. Roosevelt	Social Change
Major Construction Projects — Hoover Dam	1933	Stephen Bechtel	Social Change
Prohibition Repealed (21st Amendment)	1933	USA people	Social Change
Alcoholics Anonymous (AA)	1935	Bill Wilson	Social Change
The General Theory of Employment, Interest and Money Deficit Economics	1936	John Mayard Keynes	Social Change
Capitalism/Organized Labor	1936	Walter Reuther	Social Change
The Commonsense Book of Baby and Child Care	1946	Benjamin Spock	Social Change
African American Sports	1947	Jackie Robinson	Social Change
Universal Declaration of Human Rights	1948	Eleanor Roosevelt	Social Change
Desegregation, One Man-One Vote	1953	Supreme Court — Chief Justice Earl Warren	Social Change
McDonalds — Father of Fast Food	1955	Ray Kroc	Social Change
Beginning of the End to Racism	1955	Rosa Parks	Social Change
Marxism on Chinese Social Order	1958	Mao Zedong	Social Change
Oral Contraceptive (1923-C. Djerassi)	1960	G. Pincus/Min-Chueh Chang/J. Rock	Social Change
Power to overcome incapacity	1960	Helen Keller	Social Change
OPEC Oil Price Control	1960	OPEC	Social Change
Peace Corps of America	1961	John F. Kennedy	Social Change
The Feminine Mystique	1963	Betty Frieden	Social Change
Folk Music Re-Invented	1963	Bob Dylan	Social Change
LSD — Chemically Altered Perception	1963	Timothy Leary	Social Change
Black Power Awareness	1964	Malcom X	Social Change

CHANGE EVENTS	YEAR	CHANGE AGENT	CATEGORY
Economic Philosophy/ Knowledge Worker	1964	Peter Drucker	Social Change
Miniskirt	1965	Mary Quant	Social Change
Social Science-Fiction TV Philosopher	1966	Gene Roddenberry	Social Change
Hippies Summer of Love	1967	Haight Ashbury	Social Change
Revisions to Spirituality	1968	Eastern Religions	Social Change
Powerful Woman Politician	1969	Golda Meir	Social Change
Feminism	1970	Germain Greer	Social Change
Telethon Fund Raising for the Sick	1970	Jerry Lewis	Social Change
Economics — Use of Government Monetary Policy	1970	Paul A. Samuelson	Social Change
Roe vs. Wade — Women's Rights	1973	US Supreme Court	Social Change
Future Revealing Movies	1977	George Lucas	Social Change
Dedication to the Helpless	1979	Mother Theresa	Social Change
Government Regulation of Economy	1979	Paul Volker	Social Change
First Female Supreme Court Justice	1981	Sandra Day O'Connor	Social Change
Future Revealing Movies	1982	Stephen Spielberg	Social Change
HIV and AIDS Epidemic Scare	1983	Gene Mutation	Social Change
Revisions to Spirituality	1988	Babba Ram Daas (Richard Alpert)	Social Change
End of Marxism	1989	Mikhail Gorbachev	Social Change
Muslim Reintegration	1989	Salmon Rushdie	Social Change
Financial Failure of Marxism	1989	USSR	Social Change
Self-help Book Mania	1990	Baby Boomers	Social Change

Integration Begins in South Africa	1994	Nelson Mandela	Social Change
Over-Population	1999	India	Social Change
Mathematics of Approximation, Computers	1900	Taylor, Hermite, Hamiltonian	Technical Uncertainty
Sinking of the Titanic	1912	SS Titanic	Technical Uncertainty
Math is Uncertain — Gödel's Proof	1931	Kurt Goedel	Technical Uncertainty
Philosophy of Science — "Falsifiability"	1934	Karl Popper	Technical Uncertainty
Atomic Power Paranoia	1979	Fission Failure at Three Mile Island	Technical Uncertainty
X-rays	1901	Wilhelm G. Roentgen	Understanding the Body
Genetics and Inheritance	1902	Mendel/Fleming/De Bries/Sutton	Understanding the Body
X-ray CAT Scans	1926	Bob Ledley	Understanding the Body
Beam Magnetic Resonance	1944	Isidor Rabi	Understanding the Body
DNA ID'ed as genetic material	1944	Oswald T. Avery	Understanding the Body
Nuclear Magnetic Resonance	1946	Edward Purcell and Felix Bloch	Understanding the Body
DNA Double Helix Structure	1953	James Watson/Francis Crick	Understanding the Body
Sub-Four Minute Mile	1954	Roger Bannister	Understanding the Body
Genetic Code-DNA=>RNA=Protein	1961	Marshall W. Nirenberg/Gobind Khorana	Understanding the Body
Real Time Ultrasound Tissue Scanning	1964	Siemens	Understanding the Body
Recombinant DNA — Restriction Enzyme	1970	H. Smith and K. Wilco	Understanding the Body
Cloned Living Frog	1970	John Gurdon	Understanding the Body
Imaging-Magnetic Resonance UNY	1973	Paul Lauterbur	Understanding the Body
CAT Scanning	1979	Cornack Hounsfield	Understanding the Body
Human Genome Project	1991	J. Craig Venter-Celera/ NIH	Understanding the Body

CHANGE EVENTS	YEAR	CHANGE AGENT	CATEGORY
WWI End of Colonialism	1914	France/Germany	War
Failed Marxism Policies	1926	Joseph Stalin	War
WWII Triggered	1939	Adolf Hitler	War
Nationalism	1939	Wane of Imperialism	War
War Generals General	1944	George S. Patton	War
Marshall Plan Bretton Woods	1945	George Marshall	War
Atomic Bomb — Hiroshima	1945	Robert Oppenheimer	War
United Nations	1945	Woodrow Wilson	War
Truman Doctrine	1947	Harry S. Truman	War
Tough Minded Democracy	1951	Winston Churchill	War
Nuclear Submarine — First Strike threat ended	1954	Adm. Hyman Rickover	War
"I Have a Dream "Civil Rights	1963	Martin Luther King Jr.	War
Vietnam War — Guerilla War	1964	France, USA	War
Peaceful Protest, UC Berkeley	1964	Mario Savio	War
Gulf of Tonkin Resolution	1964	US Congress	War
Imagine — Music of the Heart	1971	John Lennon	War
Genocide	1975	Pol Pot	War

APPENDIX 3

LIST BY LAST NAME

CHANGE EVENTS	YEAR	CHANGE AGENT (Alphabetized)	CATEGORY
Post-it notes	1980	3M Co. — S. Silver/A.Fry	Life Utilities
Inside the Mind	1911	Alfred Adler	Psychology
Same Day Global Interconnection	1956	Airfreight	Mobility/Communication
Revolutionize road transportation	1929	Autobahn –Germany	Mobility/Communication
Revisions to Spirituality	1988	Babba Ram Daas-Richard Alpert	Social Change
19th Amendment — Women's Vote	1920	Cady Stanton/ Susan B. Anthony	Social Change
Prohibition — Volstead Act and18th Amendment	1920	WCTU/Anti-Saloon League	Social Change
Super Regenerative Radio Amplifier	1914	Edwin Howard Armstong	Mobility/Communication
Hippies Summer of Love	1967	Haight Ashbury	Social Change
Fiber Optic telephony (wide use 1977)	1970	AT&T	Mobility/Communication
Modernized Turkey for Muslim Women	1923	Mustafa Kemal Pasha Ataturk	Social Change
DNA ID'ed as genetic material	1944	Oswald T. Avery	Understanding the Body
Bakelite (first fully synthetic plastic)	1910	Leo Baekeland	Life Utilities
Classical American Ballet	1948	George Balanchine	Fine Art, Music and Dance
Sub-Four Minute Mile	1954	Roger Bannister	Understanding the Body
Insulin for habilitation of diabetics	1922	F. Banting, C. Best; J.J.R. McCleod, J. Collip, University of Toronto	Health
Superconductivity	1957	J. Bardeen/L.Cooper/J.Schrieffer/U ILL.	Knowledge of Universe

Heart Transplant	1967	Christian Barnard	Health
Rock and Roll world wide	1963	The Beatles	Mobility/Communication
Major Construction Projects-Hoover Dam	1933	Stephen Bechtel	Social Change
Rock and Soul	1955	Charles E. Anderson "Chuck" Berry	Mobility/Communication
IQ test	1905	A. Binet and T. Simon	Psychology
Frozen Fish Process	1923	Clarence Birdseye	Life Utilities
Ball Point Pen	1938	Laszlo Biro	Life Utilities
Negative Feedback Amplifier	1927	Harold Black	Mobility/Communication
Mathematical Encryption/Security (1918)	1945	Enigma, Bletchley Park	Mobility/Communication
Ruby Laser	1960	Theodore Maiman Bloembergen	Mobility/Communication
"Hill Street Blues" real TV drama	1981	Steven Bochco	Recording Life
Subatomic Behavior	1913	Niels Bohr	Knowledge of Universe
Radioactive Decay (Carbon) Dating	1907	Bertram Borden Boltwood	Earth History
Self-help Book Mania	1990	U.S. Boomers Born 1946-64	Psychology
Self-help Book Mania	1990	Baby Boomers	Social Change
Green Revolution — higher grain yields	1968	Norman Ernest Borlaug	Earth History
Uncertainty Physics	1927	Born/Schroedinger/Heisenberg	Knowledge of Universe
Standup TV Comedy	1930	Marx Brothers	Humor
Oil Cracking for Gasoline	1912	William Burton	Mobility/Communication
First Electrical Computer	1930	Vannevar Bush	Mobility/Communication
Large structure/sculpture, Mobile	1932	Alexander Calder	Fine Art, Music and Dance
Magnetic Tape Recording Heads	1941	Marvin Camras	Mobility/Communication

CHANGE EVENTS	YEAR	CHANGE AGENT (Alphabetized)	CATEGORY
L' Etranger (The Stranger)	1942	Albert Camus	Philosophy
Philosophy of Existentialism	1943	JP Sartre/Albert Camus	Philosophy
Xerox dry Copying	1958	Chester Carlson	Mobility/Communication
Nylon-synthetic fabrics	1937	Wallace H. Carothers	Life Utilities
Air Conditioning, temperature control etc.	1902	Willis Haviland Carrier	Life Utilities
Silent Spring	1962	Rachel Carson	Earth History
Sound Movies	1922	T.W. Case	Mobility/Communication
Neutron	1932	James Chadwick	Knowledge of Universe
Movie Comedy	1915	Charlie Chaplin	Recording Life
China Abolishes Slavery	1910	China	Social Change
Syntactic Structures	1957	Noam Chomsky	Psychology
Jazz Solo Guitar	1933	Charlie Christian / Django Reinhardt	Fine Art, Music and Dance
Tough Minded Democracy	1951	Winston Churchill	War
2001: A Space Odyssey	1968	Arthur C. Clark	Poetry and Literature
Acid Rain	1967	Coal Electric Power Generation	Earth History
Tetracycline broad antibiotic	1955	Lloyd Conover	Health
Cell Phone	1977	Cooper at Motorola, AT&T	Mobility/Communication
Fiber Optic Cable/Communication	1968	Corning Corp	Mobility/Communication
Electro-static Precipitator (removes particulates)	1923	Fred Cottrell	Life Utilities
Scuba Aqua Lung	1943	Jacques Cousteau	Life Utilities
First Artificial Heart-short lived success	1966	Michael Ellis De Bakey	Health

Wave/Particle	1923	Louis de Broglie	Knowledge of Universe
RF Amplifier Vacuum Tube, Audion, Triode	1908	Lee de Forest	Mobility/Communication
First Women's Vote	1906	Finland's Diet	Social Change
RSA public Key (Diffie-Hellman)	1976	Whitfield Diffie/Martin Hellman	Mobility/Communication
Cashless Society — Credit Card	1950	Diner's Club	Mobility/Communication
e Spin	1933	Paul Dirac	Knowledge of Universe
Disneyland	1955	Walt Disney	Fantasy
Video Tape	1956	Charles Ginsberg/Ray Dolby	Mobility/Communication
Sulfa Drugs	1935	Gerhard Domagk-IG Farben/P. Gelmo	Health
Home Refrigerator (Domestic Electric Refrigerator)	1913	Domelre/Alfred Mellowes	Life Utilities
Gyroscope (MIT)	1901	Charles Draper	Mobility/Communication
Scotch Tape	1929	Richard Drew/3M	Life Utilities
Economic Philosophy/ Knowledge Worker	1964	Peter Drucker	Social Change
"Dance of the Future"	1903	Isadora Duncan	Fine Art, Music and Dance
Folk Music Re-Invented	1963	Bob Dylan	Social Change
Color Film	1928	George Eastman	Recording Life
Invention of the Electric Utility	1903	Thomas Alva Edison	Life Utilities
Aswan High Dam	1960	Egypt	Dwellings and Structures
White rice vs. Brown (health factor)	1901	Christaan Eijkman/Gerrit Grijns	Health
Invention of light Quanta	1905	Albert Einstein-Based on Planck	Knowledge of Universe
Special Relativity $E=mC^2$	1905	Albert Einstein	Knowledge of Universe
USA Interstate Highway System	1953	President Dwight D. Eisenhower	Life Utilities

CHANGE EVENTS	YEAR	CHANGE AGENT (Alphabetized)	CATEGORY
The Hollow Men	1925	T.S. Eliot	Poetry and Literature
Public Keys, Encryption	1970	James Ellis, C Clocks	Mobility/Communication
	1973	Martin Hellman/M Williamson	
Stereophonic Sound (dual sources)	1931	A.D. English	Fine Art, Music and Dance
Cleaner Exhaust Automobiles	1970	U.S. Government, EPA Clean Air Act	Earth History
Child Development	1950	Erik Erikson	Psychology
Ragtime Respectability — Carnegie Hall	1912	James R. Europe, Clef Club Band	Fine Art, Music and Dance
Ocean Floor Topography — Plate Tectonics	1960	W. Maurice Ewing and Columbia	Earth History
Documented Ozone hole in Antarctic	1985	Farman/ Gardiner/Shanklin/UK	Earth History
Easy Graphical Transmission	1990	FAX	Mobility/Communication
Fast Mail	1985	Federal Express Co.	Mobility/Communication
Fission Reaction/Nuclear Electric Power	1939	Enrico Fermi	Life Utilities
Jazz vocal recordings	1938	Ella Fitzgerald	Fine Art, Music and Dance
Vacuum Tube – Diode/Rectifier/Detector	1904	Sir John Ambrose Fleming	Mobility/Communication
Penicillin Discovered	1928	Alexander Fleming	Health
Importance of (Vitamins) to Health	1906	W. Fletcher / F. Hopkins	Health
Penicillin Produced	1939-	H. Florey-E. Chain-Oxford, UK/	Health
	1945	A. Morey/J. Kane-Pfizer	
Automobile Assembly Line	1908	Henry Ford	Mobility/Communication
Quarks	1968	H.Freidman/H. Kendall/ R. Taylor	Knowledge of Universe
Electronic Quartz Watch	1968	Centre Electronique Horloge	Life Utilities
Interpretation of Dreams	1900	Sigmund Freud	Psychology

The Feminine Mystique	1963	Betty Frieden	Social Change
"Serpentine Dance"	1900	Loie Fuller	Fine Art, Music and Dance
Vitamin(e) Chemistry — an Amine	1911	C. Funk	Health
Hologram Laser 3D Images	1947	Dennis Gabon	Knowledge of Universe
First Personed Satellite	1961	USSR, Gagarin	Mobility/Communication
Gallup Polls- Reliable Polling	1935	George Gallup	Mobility/Communication
Cosmic Background Radiation	1964	G. Gamow (1916 Dicke)	Knowledge of Universe
Progressive Civil Disobedience	1920	Mohandas K. Gandhi	Social Change
Dr. Seuss, Green Eggs and Ham	1955	Theodore Seuss Geisel	Learn and Teach Tools
HIV and AIDS Epidemic Scare	1983	Gene Mutation	Social Change
WWI End of Colonialism	1914	France/Germany	War
Classical Music — Musical Theater	1928	George Gershwin	Fine Art, Music and Dance
Safety Razor	1901	King Camp Gillette	Life Utilities
"Rocket Reaching Extreme Altitudes" (V2= 1944)	1926	Robert H. Goddard	Mobility/Communication
Math is Uncertain — Gödel's Proof	1931	Kurt Goedel	Technical Uncertainty
The Lord of the Flies	1954	William Golding	Poetry and Literature
The LP Vinyl Record	1948	Peter Goldmark	Fine Art, Music and Dance
Chimpanzee Tool Use (similar to human)	1960	Jane Goodall	Earth History
End of Marxism	1989	Mikhail Gorbachev	Social Change
Laser/Maser	1958	Gould, Townes and Schawlow	Mobility/Communication
"Lamentation" "Appalachian Spring"(1942)	1930	Martha Graham and Aaron Copland	Fine Art, Music and Dance
Implantable Pacemaker	1958	W. Greatbatch and W. Chardack	Health

CHANGE EVENTS	YEAR	CHANGE AGENT (Alphabatized)	CATEGORY
Feminism	1970	Germain Greer	Social Change
Functional Architecture	1924	Walter Gropius	Dwellings and Structures
Cloned Living Frog	1970	John Gurdon	Understanding the Body
American Folk Music	1937	Woody Guthrie	Fine Art, Music and Dance
First Synthetic Vitamin — Vitamin C	1933	Szent-Gyorgyi/WaughandKing/T. Reichistein, Swiss Inst. Tech./C. Haworth-UK Birmingham	Health
Fission	1939	Hahn/Meitner; Strassman/Bohr	Knowledge of Universe
Laser and Light Emitting Diode (LED)	1962	Robert Hall/ Nick Holonyak Jr.	Mobility/Communication
Stage Musicals	1943	Oscar Hammerstein	Recording Life
A Brief History of Time	1991	Stephen Hawking	Poetry and Literature
External Heart Pacemaker (unsuccessful)	1932	A.S. Hyman	Health
Matrix Mechanics	1925	Werner Heisenberg	Knowledge of Universe
For Whom the Bell Tolls (We are one nation)	1929	Ernest Hemingway	Mobility/Communication
Muppets — Puppet Educators	1969	Jim Henson	Learn and Teach Tools
World War II Triggered	1939	Adolf Hitler	War
Software — Nanosecond Wire	1945	Adm. Grace Hopper	Learn and Teach Tools
External Heart Pacemaker	1951	J. A. Hopps-Banting Institute	Health
CAT Scanning	1979	Cornack Hounsfield	Understanding the Body
Father of Blues	1927	Son House	Fine Art, Music and Dance
Universe Bigger Than Milky Way	1924	Edwin Hubbel	Knowledge of Universe
Blood Bank	1914	Albert Hustin	Health

CHANGE EVENTS	YEAR	CHANGE AGENT (Alphabetized)	CATEGORY
Child/Adult Moral Development	1968	Lawrence Kohlberg	Psychology
McDonalds — Father of Fast Food	1955	Ray Kroc	Social Change
New York Armory Show of Modern Art	1913	W. Kuhn, A. B. Davies	Fine Art, Music and Dance
Spread Spectrum (idea, later used for cell phone)	1942	Hedy Lamarr and George Antheil	Mobility/Communication
Polaroid Film	1932	Edwin H. Land	Recording Life
Imaging-Magnetic Resonance	1973	Paul Lauterbur, UNY	Understanding the Body
Cyclotron	1930	Ernesto Lawrence	Knowledge of Universe
The Skyscrapers	1923	Charles E. Jeanneret/Le Corbusier	Dwellings and Structures
Australopithecus — Homo Habilus	1961	Louis Seymour Bazett Leakey	Earth History
"All in the Family" Real People	1971	Norman Lear	Recording Life
LSD — Chemically Altered Perception	1963	Timothy Leary	Social Change
X-ray CAT Scans	1926	Bob Ledley	Understanding the Body
To Kill a Mockingbird	1961	Harper Lee	Poetry and Literature
Bolshevik Revolt	1917	Vladimir Lenin	Social Change
"Imagine" — Music of the Heart	1971	John Lennon	War
Telethon Fund Raising for the Sick	1970	Jerry Lewis	Social Change
Leads on Heart Pacemaker (w/ ext. battery)	1955	C. W. Lillehei and E. Bakken	Health
Radio Astronomy	1957	Bernard Lovell	Knowledge of Universe
Future Revealing Movies	1977	George Lucas	Social Change
Integration begins in South Africa	1994	Nelson Mandela	Social Change
Transatlantic Wireless Telegraph	1901	Guglielmo Marconi/Karl F. Braun	Mobility/Communication

Quartz Clock — Accurate Time Keeping	1927	W.A. Marrison and J.W. Horton	Mobility/Communication
Marshall Plan Bretton Woods	1945	George Marshall	War
ENIAC Computer	1946	Mauchly/ Eckert /John /J. Presper Jr.	Learn and Teach Tools
First Vitamin (A) Isolated	1913	Davis and McCollum, University of Wisconsin/ Osborn and Mendel, Yale	Health
Container Ship (Ideal-X)	1956	Malcom Mclean	Mobility/Communication
Cyclotron for Physics Studies	1945	E.M. McMillan/V. Veksler	Knowledge of Universe
Powerful Woman Politician	1969	Golda Meir	Social Change
Genetics and Inheritance	1902	Mendel/Fleming/De Bries/Sutton	Understanding the Body
Movies: MGM	1917	Louis Meyer	Recording Life
Historical Novel	1948	James Albert Michener	Learn and Teach Tools
CFC = Chlorofluorohydrocarbons (noncorrosive)	1930	Thomas Midgley Jr.	Life Utilities
Velcro (idea from a plant) (hook and loop fastener)	1948	George de Mestral	Life Utilities
Ozone Layer Impact of Freon	1974	Molina, Crutzen,Rowland–UCI	Earth History
Montessori Method	1906	Maria Montessori	Learn and Teach Tools
High Temperature Superconductors	1964	K.A. Müller/J.D. Bednorz	Knowledge of Universe
DDT= Insect/Typhus (disease control pesticide)	1939	Muller/Ziedler	Life Utilities
Kidney Transplant	1954	Joseph E. Murray	Health
First Communications Satellite Telstar I	1962	AT&T/NASA/Bell Labs	Mobility/Communication
AIDS Cocktail (to slow the effects)	1996	NIH — many contributors	Health
Genetic Code — DNA=>RNA=Protein	1961	Marshall W. Nirenberg/ Gobind Khorana	Understanding the Body

CHANGE EVENTS	YEAR	CHANGE AGENT (Alphabetized)	CATEGORY
Integrated Circuit	1959	Robert Noyce/Jack Kilby	Mobility/Communication
First Female Supreme Court Justice	1981	Sandra Day O'Connor	Social Change
Creole Jazz Band	1923	King Oliver, Jelly Roll Morton	Fine Art, Music and Dance
Plays of American Severity	1936	Eugene O'Niell	Recording Life
Superconductivity	1911	H.K. Onnes	Knowledge of Universe
OPEC Oil Price Control	1960	OPEC	Social Change
Atomic Bomb — Hiroshima	1945	Robert Oppenheimer	War
Animal Farm	1948	George Orwell	Poetry and Literature
1984	1945	George Orwell	
Pap Smear Test	1943	G. Papanicolaou, H.Traut, Columbia University	Health
Beginning of the End to Racism	1955	Rosa Parks	Social Change
War Generals General	1944	George S. Patton	War
Opera	1965	Luciano Pavarotti	Fine Art, Music and Dance
The Power of Positive Thinking	1952	Norman V. Peale	Psychology
Zone Refining of Silicon	1952	William Pfann	Mobility/Communication
Child Development Phases	1929	Jean Piaget	Psychology
Modern Art	1907	Pablo Picasso (plus hundreds of artists)	Fine Art, Music and Dance
Oral Contraceptive (C. Djerassi, 1923)	1960	G. Pincus/Min-Chueh Chang/J. Rock	Social Change
Quantum Theory	1900	Max Planck	Knowledge of Universe
Teflon-PTFE Neutral High-temperature Plastic	1938	Roy Plunkett	Life Utilities

Philosophy of Science — "Falsifiability"	1934	Karl Popper	Technical Uncertainty
Genocide	1975	Pol Pot	War
Magnetic Wire Recording (V. Poulsen 1898)	1928	J. Begun/ F. Pfleumer	Mobility/Communication
Rock and Roll	1954	Elvis Presley	Fine Art, Music and Dance
Nuclear Magnetic Resonance	1946	Edward Purcell and Felix Bloch	Understanding the Body
Miniskirt	1965	Mary Quant	Social Change
Beam Magnetic Resonance	1944	Isidor Rabi	Understanding the Body
Individual Objectivism	1957	Ayn Rand	Philosophy
First Congresswoman	1917	Jeannette Rankin	Social Change
Color TV	1953	RCA (NTSC and Shadow Mask)	Mobility/Communication
Liquid Crystal Display (LCD)	1971	RCA, Kent State University, LaRoche Corporation	Life Utilities
Revisions to Spirituality	1968	Eastern Religions	Social Change
Capitalism/Organized Labor	1936	Walter Reuther	Social Change
Nuclear Submarine (first strike threat ended)	1954	Adm. Hyman Rickover	War
African American Sports	1947	Jackie Robinson	Social Change
American Ideal Family Paintings	1930	Norman Rockwell	Recording Life
Social Science-fiction TV Philosopher	1966	Gene Roddenberry	Social Change
X-rays	1901	Wilhelm G. Roentgen	Understanding the Body
Composite Carbon-fiber Material Strength-Weight	1950s	Rolls Royce/Royal Air Force	Life Utilities
Entitlement, Welfare, The New Deal	1933	Franklin D. Roosevelt	Social Change
Universal Declaration of Human Rights	1948	Eleanor Roosevelt	Social Change

CHANGE EVENTS	YEAR	CHANGE AGENT (Alphabatized)	CATEGORY
Non-Local Realism	1933	Puldansky, Rosen, Einstein	Knowledge of Universe
Offset Printing Becomes Real	1904	Ira Rubel	Mobility/Communication
Offset Printing Becomes Real	1904	Ira Rubel	Poetry and Literature
Muslim Reintegration	1989	Salmon Rushdie	Social Change
Interesting Problems	1910	Bertrand Russell	Philosophy
Star Lives — The Main Sequence	1914	Henry Russell	Knowledge of Universe
Sports Hero Model	1927	Babe Ruth	Social Change
Atomic Nucleus	1911	Ernest Rutherford	Knowledge of Universe
Proton	1914	Ernest Rutherford	Knowledge of Universe
Le Petit Prince (Friendship and hope defined)	1943	Antoine de Saint Exupéry	Philosophy
Salk Vaccine for Polio	1952	Jonas Salk	Health
Economics — Use of Government Monetary Policy	1970	Paul A. Samuelson	Social Change
First American Birth Control Clinic (Becomes Planned Parenthood in 1942)	1915	Margaret Sanger	Social Change
TV	1939	David Sarnoff	Recording Life
Peaceful Protest, UC Berkeley	1964	Mario Savio	War
12-tone Serial Technique	1921	Arnold Schoenberg	Fine Art, Music and Dance
Music and Medicine for the Poor	1920	Albert Schweitzer	Social Change
Popular American Folk Music	1949	Pete Seeger	Fine Art, Music and Dance
Classical Guitar	1920	Andres Segovia	Fine Art, Music and Dance
Anti-proton	1955	E. G. Segre/O. Chamberlain	Knowledge of Universe

American Slash Poetry	1960 Anne Sexton	Poetry and Literature
India tonality Influence on Rock via The Beatles	1956 Ravi Shankar	Fine Art, Music and Dance
Information Theory	1949 C.E. Shannon	Knowledge of Universe
The Transistor	1948 Brattain, Bardeen, Shockley	Mobility/Communication
Real Time Ultrasound Tissue Scanning	1964 Siemens	Understanding the Body
Helicopters — Vertical Lift	1939 Igor Sikorsky	Mobility/Communication
The Jungle	1906 Upton Sinclair	Poetry and Literature
Cheap High Quality CD Music Reproduction	1979 Joop Sinjou & Toshitada Doi	Mobility/Communication
Skinner Box Modifies Behavior	1938 Burrhus Frederic Skinner	Psychology
Recombinant DNA — Restriction Enzyme	1970 H. Smith & K. Wilco	Understanding the Body
Twenty to Forty Million Influenza Deaths	1918 Swine (Spanish) Flu	Health
Microwave Oven	1947 Percey L. Spencer	Life Utilities
Future Revealing Movies	1982 Stephen Spielberg	Social Change
Global Positioning System	1987 J. J. Spilkher	Mobility/Communication
The Commonsense Book of Baby and Child Care	1946 Benjamin Spock	Social Change
Failed Marxism Policies	1926 Joseph Stalin	War
East of Eden	1952 John Steinbeck	Poetry and Literature
Black and White Photography as Art	1902 Alfred Stieglitz	Recording Life
Atomic Power Paranoia	1979 Fission Failure at Three Mile Island	Technical Uncertainty
Comedy Slapstick	1934 Three Stooges	Humor
Mathematics of Approximation, Computers	1900 Taylor, Hermite, Hamiltonian	Technical Uncertainty
Pocket Calculator	1971 Texas Instruments Inc.	Life Utilities

CHANGE EVENTS	YEAR	CHANGE AGENT (Alphabetized)	CATEGORY
Dedication to the Helpless	1979	Mother Theresa	Social Change
Americanization of English Humor	1925	James Thurber	Humor
Sinking of the Titanic	1912	SS Titanic	Technical Uncertainty
CAD (computer aided design) Software	1982	Many contibutors (see Chapter 4 reference notes)	Life Utilities
The Rise and Fall of Cultures	1934	Arnold J. Toynbee	Recording Life
Airlines — Pan American	1935	John Trip	Mobility/Communication
Truman Doctrine	1947	Harry S. Truman	War
Automotive Smog Defined — Los Angeles	1946	R. R. Tucker/Haagen-Smit	Earth History
Computer Programs/Artificial Intelligence	1937	Alan M. Turing	Learn and Teach Tools
Computers/programs/ Artificial Intelligence	1937	Alan M. Turing	Life Utilities
International Geophysical Year — Atmospheric Monitoring	1957	World governments	Earth History
Montreal Protocol — Limit Atmospheric CFC's	1987	United Nations & scientists	Earth History
Kyoto Protocol — Limit CO2 (Greenhouse Gas)	1996	UNFCCC	Earth History
16th Amendment	1913	Personal Income Tax	Government
Gulf of Tonkin Resolution	1964	US Congress	War
Roe vs. Wade — Womens Rights	1973	US Supreme Court	Social Change
Panama Canal	1914	United States	Dwellings and Structures
Home Refrigerator (Domestic Electric Refrigerator)	1913	Domeire by F. Wolf /A. Mellowes	Life Utilities
U.S. Stock Market Crash-American Depression	1929	Wall Street USA	Social Change
Gold Standard for Currency Abolished	1933	US Government	Life Utilities

Prohibition Repealed (21st Amendment)	1933	USA people	Social Change
First Rocket Reaches the Moon	1959	USA	Knowledge of Universe
First Person on the Moon	1969	USA Kennedy	Knowledge of Universe
Volstead Act Repealed Prohibition Ended	1933	Democratic moralism	Government
First Satellite – Sputnik	1957	USSR	Knowledge of Universe
First Rocket Landing on Mars	1971	USSR	Knowledge of Universe
Nuclear Reactor Disaster — Chernobyl	1986	USSR	Life Utilities
Financial Failure of Marxism	1989	USSR	Social Change
Human Genome Project	1991	J. Craig Venter, Celera/ NIH	Understanding the Body
24-gear Mountain Bikes	1910-1987	P. de Vivie/T. Campagnolo/Tour de France	Life Utilities
Government Regulation of Economy	1979	Paul Volker	Social Change
Computer Architecture	1946	John von Neumann	Learn & Teach Tools
Real Sound Movies — The Jazz Singer	1927	Warner Brothers	Mobility/Communication
Desegregation, One Man-One Vote	1953	Supreme Court — Chief Justice Earl Warren	Social Change
Wassermann Test for Syphilis	1906	August von Wassermann	Health
Jazz/Blues	1943	Muddy Watters	Fine Art, Music and Dance
Beginnings of Behaviorist Theory	1913	John B. Watson	Psychology
Behavioral Psychology	1914	John Broadus Watson	Psychology
DNA Double Helix Structure	1953	James Watson/Francis Crick	Understanding the Body
Origins of Continents & Oceans	1915	Alfred Lothar Wegener	Earth History
First Jet Engine Aircraft	1941	Sir Frank Whittle	Mobility/Communication

CHANGE EVENTS	YEAR	CHANGE AGENT (Alphabetized)	CATEGORY
Sedatives	1952	Robert Wilkins	Life Utilities
Alcoholics Anonymous (AA)	1935	Bill Wilson	Social Change
United Nations	1945	Woodrow Wilson	War
Tractates Logico-Philosophicus	1920	Ludwig Wittgenstein	Philosophy
Chlorination of Water — Purification	1915	Abel Wolmand	Health
Apple II	1977	Seve Wozniak, Steven P. Jobs	Mobility/Communication
Powered Flight	1903	Wilbur and Orville Wright	Mobility/Communication
American Architecture	1909	Frank Lloyd Wright	Dwellings and Structures
World Wide Web (WWW)	1989	DARPA, UCLA, CERN, MIT-Mosaic, HTTP, HTML, TCP/IP, DNS	Mobility/Communication
Middle America Realism Painting	1937	Andrew Wyeth	Fine Art, Music and Dance
Black Power Awareness	1964	Malcom X	Social Change
Transition Poetry	1900	William Butler Yeats	Poetry and Literature
Marxism on Chinese Social Order	1958	Mao Zedong	Social Change
Electro-magnetic Waves	1910	Pieter Zeeman/ Hendrick A. Lorentz	Mobility/Communication
External Portable Resuscitator	1952	P. Zoll- Beth Israel Hospital	Health
TV Camera Tube (orthicon, vidicon)	1923	Vladimir Kosma Zworykin	Mobility/Communication
Cathode Ray Tube (CRT)	1939	Zworykin, Farnsworth, Dumont	Mobility/Communication

APPENDIX 4

PHOTO CREDITS

CHAPTER 0
Page 2, New York: Photo Alfred Stieglitz, 1893, National Gallery of Art, Washington D.C., Alfred Stieglitz Collection
Page 6, Combine: Photo AGCO Corporation/Massey Ferguson

CHAPTER 1
Page 15, Suffrigists: Photo W. R. Ross, Prints and Photographs Division, Library of Congress, LC-USZ62-8312.
Page 16, Hotel Clark: Marion Post Wolcott, Prints and Photographs Division, Library of Congress, LC-USF33-30637-M3
Page 18, Women welders: Photo Alfred T Palmer, Prints and Photographs Division, Library of Congress, LC-USF33-30637-M3.
Page 21, Birth control: Photo Tina Sbrigato, iStockPhoto

CHAPTER 2
Page 28, Freeway Intersection: Terraserver.com
Page 35, Tube radio: Photo David Coder, iStockPhoto
Page 37, Container ship: Photo David Partington, iStockPhoto
Page 38, DC-3: Photo Michael Carter, http:www.jetphotos.net

CHAPTER 3
Page 46, Saturn V: NASA
Page 47, Galaxy: NASA
Page 51, Earth: U.S. National Oceanic and Atmospheric Administration, NGDC-ETOPO2
Page 55, Saturn V: NASA

CHAPTER 4
Page 69, Energy use: 2003 Annual Energy Review, DOE/EIA-0384(2003) Source: Energy Information Administration, www.eia.doe.gov/aer
Page 70, Renewable energy: 2003 Annual Energy Review, DOE/EIA-0384(2003) Source: Energy Information Administration, www.eia.doe.gov/aer
Page 74, Energy demand: 2003 Annual Energy Review, DOE/EIA-0384(2003) Source: Energy Information Administration, www.eia.doe.gov/aer
Page 75, Earth light: NASA
Page 76, Energy consumption: DOE/EIA-0384(2003) Source: Energy Information Administration, www.eia.doe.gov/aer

CHAPTER 5
Page 86, MRI: Siemens Medical Solutions, Ultrasound Division
Page 87, Heart image: Siemens Medical Solutions, Ultrasound Division
Page 91, Survival curve: National Vital Statistics Reports, Vol. 53, No. 6, Nov. 10, 2004 provided on the Center for Disease Control website, www.cdc.gov/nchs/nvss.htm

CHAPTER 6

Page 101, Walking Man II: Hirshhorn Museum and Sculpture Garden, Smithsonian Institution, Gift of Joseph H. Hirshhorn, 1966. Photo Lee Stalsworth
Page 102, Guernica: Pablo Picasso (1881-1973) ç ARS, NY, Museo Nacional Centro de Arte Reina Sofia, Madrid, Spain. John Bigelow Taylor, Art Resource, NY
Page 106, Loie Fuller sculpture: Raoul Larche, sculpture; Ophir Gallery

CHAPTER 8

Page 137, Panama Canal: Photo F.R. Roberson, Prints and Photographs Division, Library of Congress, LC-USZ62-117339
Page 138, Hong Kong: Photo Adam Korzekwa, iStockPhoto
Page 139, Spending: Congressional Joint Economic Committee Study 1998
Page 142, Cemetery: Photo Pietro Valdinoci, iStockPhoto

SELECTED BIBLIOGRAPHY

Clark, Kenneth. *Civilisation: A Personal View,* Harper & Row, NY, 1969
Daniel, Clifton; Schlesinger, A. M. *20th Century Day by Day,* Doorling & Kindersley 1992, NY, 2004
Fernandez-Armesto, Felipe. *Ideas that changed the world,* Doorling Kindersley,DK Publishing NY, 2003
Gilbert, Martin. *History of the 20th Century,* (three volume version and a concise version) Harper Collins Inc./Perennial, NY, 2001, 300 illustrations, 36maps, 1000 pages
Grenville, J.A.S. *History of the World in the 20th Century.* Belknap Harvard Press, 84, 2000
Grun, Bernard. *The Timetables of History,* Simon & Schuster/Touchestone, NY, 1975, (based on Stein's KulturFahrPlan below)
Howard, Michael; Louis, Wm Roger. *The Oxford History of the Twentieth Century,* Oxford University Press, NY, 1998
Murray, Charles. *Human Accomplishment,* Harper Collins, NY, 2003
Roberts, J. M. *The History of the World 1901-2000,* Penguin Books, NY, 1999
Stein,Werner. *KulturFahrPlan,* F A Herbig Verlagsbuchhandlung, Germany, 1946
Stolley, R. B. *Life — Our Century in Pictures,* Little, Brown and Company/Bulfinch Press, NY, 2000
Teeple, John B. *Timelines of World History,* Doorling Kindersley, DK Publishing, NY, 2002
Time/CBS News. *People of the Century,* Simon & Schuster, NY, 1999
Watson, Peter. *The Modern Mind- An Intellectual History of the 20th Century,* Harper Collins/Perennial, NY, 2001

READER'S LIST PAGES

These pages are for you, the reader, whose ideas and opinions deserve their place on the lists.

CHANGE EVENTS	**CHANGE AGENT**

CHANGE EVENTS   **CHANGE AGENT**

CHANGE EVENTS	CHANGE AGENT

6